Lecture Notes in Electrical Engineering

Volume 568

The book series *Lecture Notes in Electrical Engineering* (LNEE) publishes the latest developments in Electrical Engineering - quickly, informally and in high quality. While original research reported in proceedings and monographs has traditionally formed the core of LNEE, we also encourage authors to submit books devoted to supporting student education and professional training in the various fields and applications areas of electrical engineering. The series cover classical and emerging topics concerning:

- Communication Engineering, Information Theory and Networks
- Electronics Engineering and Microelectronics
- Signal, Image and Speech Processing
- Wireless and Mobile Communication
- Circuits and Systems
- Energy Systems, Power Electronics and Electrical Machines
- Electro-optical Engineering
- Instrumentation Engineering
- Avionics Engineering
- Control Systems
- Internet-of-Things and Cybersecurity
- Biomedical Devices, MEMS and NEMS

For general information about this book series, comments or suggestions, please contact leontina.dicecco@springer.com.

To submit a proposal or request further information, please contact the Publishing Editor in your country:

China

Jasmine Dou, Associate Editor (jasmine.dou@springer.com)

India

Swati Meherishi, Executive Editor (swati.meherishi@springer.com)
Aninda Bose, Senior Editor (aninda.bose@springer.com)

Japan

Takeyuki Yonezawa, Editorial Director (takeyuki.yonezawa@springer.com)

South Korea

Smith (Ahram) Chae, Editor (smith.chae@springer.com)

Southeast Asia

Ramesh Nath Premnath, Editor (ramesh.premnath@springer.com)

USA, Canada:

Michael Luby, Senior Editor (michael.luby@springer.com)

All other Countries:

Leontina Di Cecco, Senior Editor (leontina.dicecco@springer.com)
Christoph Baumann, Executive Editor (christoph.baumann@springer.com)

**** Indexing: The books of this series are submitted to ISI Proceedings, EI-Compendex, SCOPUS, MetaPress, Web of Science and Springerlink ****

More information about this series at http://www.springer.com/series/7818

Wei Li · Shengchen Li · Xi Shao · Zijin Li
Editors

Proceedings of the 6th Conference on Sound and Music Technology (CSMT)

Revised Selected Papers

Springer

Editors
Wei Li
School of Computer Science
and Technology
Fudan University
Shanghai, China

Xi Shao
Institute of Telecommunications
and Information Engineering
Nanjing University of Posts
and Telecommunications
Nanjing, Jiangsu, China

Shengchen Li
Beijing University of Posts
and Telecommunications
Beijing, China

Zijin Li
China Conservatory of Music
Beijing, China

ISSN 1876-1100 ISSN 1876-1119 (electronic)
Lecture Notes in Electrical Engineering
ISBN 978-981-13-8706-7 ISBN 978-981-13-8707-4 (eBook)
https://doi.org/10.1007/978-981-13-8707-4

This Springer imprint is published by the registered company Springer Nature Singapore Pte Ltd.
The registered company address is: 152 Beach Road, #21-01/04 Gateway East, Singapore 189721, Singapore

Preface

For the first time ever, the leading Chinese computer audition conference, the Conference on Sound and Music Technology (CSMT), has published separate proceedings in English. This remarkable event is a perfect present for the sixth birthday of CSMT and is a great opportunity for Chinese machine audition researchers to present their ideas to the world as an organised group.

The predecessor of CSMT was the China Sound and Music Computing Workshop (CSMCW), which was initially organised by Tsinghua University and Fudan University in 2013. Another main organiser of the conference, the Shanghai Computer Music Association, has been involved in running the conferences since 2014. The first two editions of CSMCW were held in Fudan University and Tsinghua University in Shanghai and Beijing, respectively, and were followed by the third edition of CSMCW hosted by Shanghai Conservatory.

In 2015, CSMCW was rebranded as CSMT since the original workshop had expanded to become an academic conference. Held by Nanjing University of Posts and Telecommunications, the fourth edition of CSMT called for papers in Chinese and the accepted papers were recommended to the Journal of Fudan for publication. The fifth edition of CSMT was a satellite workshop of the world-leading computer music conference, ISMIR (International Society of Music Information Retrieval), which was held by Soochow University, Suzhou, China.

The sixth edition of CSMT was held in Xiamen, China. This was the first CSMT conference calling for English submissions and the first to publish its own proceedings. Since 2013, the number of attendees has grown rapidly, from a couple of dozen to more than 200.

Despite being well recognised as an independent research domain for a decade, machine audition, which encompasses computer-based music processing and analysis and acoustic signal processing, has not been a separate research domain in China but has been considered as a part of traditional research domains, such as multimedia signal processing, automation, musicology and audio engineering.

As the organisers of CSMT, we believe the growth of CSMT will help machine audition to become an independent research domain in China and machine audition will play an important role in the rapid development of China. Moreover, with the development of machine audition all over the world, researchers in China can make better use of the existing experience by making it an independent research domain, such as stronger collaborations between musicologists, psychologists and engineers, and embracing acoustic signal processing as an important part of the field.

In conclusion, this milestone publication of proceedings in English will stimulate the growth of computer audition as a separate research domain.

Shanghai, China	Wei Li
Beijing, China	Shengchen Li
Nanjing, China	Xi Shao
Beijing, China	Zijin Li

Contents

Music Steganography

Music Processing and Music Information Retrieval

method achieves 16% improvement in accuracy compared with the state-of-the-art singer identification system.

Keywords Singer identification (SI) · Universal background model (UBM) · Gaussian mixture model (GMM)

1 Introduction

Music is an indivisible part of our lives for recreational purposes. The singing voice is one of the most intriguing and informative attributes of the music [1]. From the perspective of music information retrieval (MIR), singer identification (SI) is a task concerning the automatic recognition of the singers of given songs.

Various studies [2–6] have been presented to model a singing voice, and vocal separation and timbral acoustic features are the two most important parts within it. Among various features, Mel-frequency Cepstral Coefficients (MFCCs) are the most popular ones [7], and Logan verified that MFCC feature can be used on music modeling [8]. In [7, 9–11], special efforts were made to distinguish the singing voice from instrumental sounds in a song. Especially, based on the structure of music, Huang [5] proposed robust principal component analysis (RPCA) for singing voice separation from musical accompaniment.

In a Gaussian mixture model (GMM) [12], Hamid used i-vector to extract song-level features and used Probabilistic Linear Discriminant Analysis (PLDA) to predict the artist for given songs from Artist20 [13]. Further, state-of-the-art GMM-based [14] speaker recognition systems used the universal background model. In this paper, we investigate the SI task, using a novel approach based on UBM.

The rest of the paper is organized as follows: Sect. 2 presents the proposed approach, Sect. 3 provides the dataset description and experimental setup, Sect. 4 reports the experimental results, and Sect. 5 concludes the paper.

2 Proposed Approach

We introduce the GMM-UBM for automatic singer identification, which is a model usually used in speaker identification. At the front end, MFCC features were extracted, after the vocal separation, as the singer vocal timbral feature. The core component of the singer identification system is the back end, where singers' voices are modeled (enrolled) and verification trials are scored. The enrollment phase includes estimating a model that represents the acoustic field of each singer.

2.1 Singer Identification System

We note S_{f_n} the short-time Fourier transform (STFT) of the input signal, and $n = [1 : N]$ being respectively frequency and time indices. S_{f_n} is the sum of singing voices V_{f_n} and A_{f_n} is the accompaniment, expressed by the following vector formula:

$$S_n = V_n + A_n \tag{1}$$

where $S_n = \left[S_{1_n}, \ldots, S_{f_n}\right]^T$, $V_n = \left[V_{1_n}, \ldots, V_{f_n}\right]^T$ and $A_n = \left[A_{1_n}, \ldots, A_{f_n}\right]^T$.

Basic K-singer identification system consists of the following two steps:

1. Modeling: For each singer P_k, where $K = 1, \ldots, K$, A GMM is derived from the feature vector x, using the maximum likelihood (MLE) method for the training songs of P_k.
2. Identification: The likelihood values of the test song are calculated with each singer model, and the label of maximum likelihood is assigned to the test song.

2.2 GMM-UBM Singer Identification

The novelty of our proposed approach lies in using GMM-UBM on singer identification. We use all the singers' songs to build a massive universal background model (UBM) for all the singers. For each singer, we train a GMM model on the UBM. Note that to get more attention on the singer's voice, we do a vocal separation as the preprocess step of all the songs. We choose the MFCC as the timbral feature. The singer identification system is shown in Fig. 1.

Our singer identification system comprises of four building blocks: (1) vocal separation, (2) feature extraction, (3) model generation, and (4) singer recognition. Details are as follows.

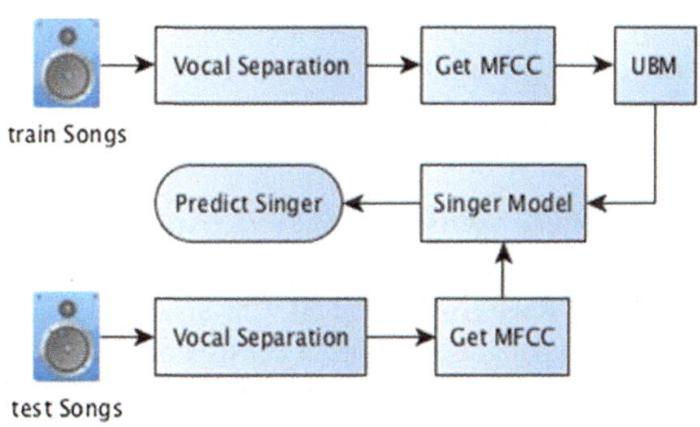

Fig. 1 Proposed singer identification system with GMM-UBM

First, all the songs in the dataset are preprocessed by vocal separation, using robust principal component analysis. This allows us to obtain the singing voice from the mixture music.

Second, 13-dimensional MFCC vectors are extracted from every segment for each track of singer P_k and, then, concatenated into a feature matrix $M_{i,j}$. Subsequently, a mean removal technique is applied on every row of matrix M, yielding a new matrix denoted M^*.

Then, the dataset is split into a training set and testing set. On the training set, we bring all the singer's song together to build a GMM supervector as UBM. The speaker-specific models are, then, adapted from the UBM, using the maximum a posteriori (MAP) estimation.

Finally, the probability of song from the testing set is computed with each singer model, and an identified singer name with maximum probability among K singers is returned.

2.3 Vocal Separation

The approach RPCA suggests solving the following convex optimization problem:

1. Minimize $\|A\|_* + \delta \|V\|_1$
2. Subject to $A + V = S$

where S $\in R^{n_1 \times n_2}$, V $\in R^{n_1 \times n_2}$,$\|\cdot\|_*$ and $\|\cdot\|_1$ denotes the nuclear norm (sum of singular values) and the L1-norm (sum of absolute values of matrix entries) respectively. δ is a trade-off parameter between the rank of A and the sparsity of V. As Candes suggested in [15], $\delta = 1/\sqrt{\max(n_1, n_2)}$.

Since music instruments can reproduce the same sounds for times and since instrumental music has an underlying repeating musical structure [5], we note the input music signal as a low-rank signal A. Singing voices, on the contrary, have more variations (higher rank) but are relatively sparse in the time and frequency domains. Therefore, we can note singing voices as components that make up the sparse matrix V. By RPCA, the input music signal S is separated into musical accompaniment A and vocal signals (the sparse matrix) V.

2.4 Model Generation

Universal Background Model. Universal backboard model (UBM) aims to model the overall data distribution. It consists of a classical GMM. For a N-dimensional feature vector x, the mixture density used for the likelihood function is defined as a weighted sum of unimodal Gaussian models:

$$p(x|\lambda) = \sum_{i=1}^{L} \omega_i p_i(x) \tag{2}$$

where M is the number of Gaussian components, $p_i \sim N(\mu_i, \sigma_i)$. λ represents the GMM parameters, where $\lambda_i = \{\omega_i, \mu_i, \sigma_i\}$, $i = 1, \ldots, L$. x represents a feature vector.

The UBM is usually composed of Gaussian models with diagonal covariance matrix, and we use 32 components in this work. The loss of modeling ability, caused by the diagonal covariance matrix, can be compensated by increasing the Gaussian number in the mixture [16]. The UBM is trained using a large and representative dataset, by using the expectation maximization (EM) algorithm. We use a maximum a posteriori (MAP) approach to parameterize the UBM. The first step is to determine the probabilistic alignment of the training vectors with the UBM Gaussian components. And, for a Gaussian G_i in the UBM, we compute follows:

$$Pr(i, x_t) = \frac{\omega_i, p_i x_t}{\sum_{j=1}^{M} \omega_j, p_j x_t} \tag{3}$$

$$n_i = \sum_{t=1}^{T} \Pr(i, x_t) \tag{4}$$

$$E_i(x) = \frac{1}{n_i} \sum_{t=1}^{T} \Pr(i, x_t) x_t \tag{5}$$

These statistical values are, then, used for adapting the mean vector $\hat{\mu}$ of each Gaussian during the following iterative process:

$$\widehat{\mu_i}^0 = \mu_i \tag{6}$$

$$\widehat{\mu_i}^k = \alpha_i E_i(x) + (1 - \alpha_i)\widehat{\mu_i}^{k-1} \tag{7}$$

$$\alpha_i = \frac{n_i}{n_i + r} \tag{8}$$

where x_t represents the tth feature vector of the music track to mode, t $= 1, \ldots,$ T, while r is a fixed constant, α_i is the iteration factor. K $= 1, \ldots,$ K represents the iteration number.

Singer Model on UBM. Each singer model is directly derived from a generic GMM, estimated by a large set of representative data—i.e., the UBM. The UBM adaptation is the process of modifying the parameters to fit a particular data distribution. During the adaption process, the mean vectors of the Gaussians are modified to fit the particular singer's song track distribution. A particular singer model can be summarized by the UBM parameters:

$$\psi = (\mu_1, \mu_2, \ldots, \mu_N)^T \tag{9}$$

where ψ, named the GMM supervector, is the concatenation of all the mean vectors of the N Gaussian components.

2.5 Model Generation

We use Kullback-Leibler (KL) divergence to approximate two models α and b by:

$$D\left(\mu^{a}, \mu^{b}\right) = \frac{1}{2}\sum_{i=1}^{n}\gamma_i\left(\mu_i^{\alpha} - \mu_i^{b}\right)^T \sum_{i}^{-1}\left(\mu_i^{\alpha} - \mu_i^{b}\right) \tag{10}$$

where μ^a and μ^b are the GMM supervectors of the models a and b respectively. γ_i represents the mixture weights and $\sum_i$ represents the covariance matrix of the Gaussian component. From this representation, we can deduce the following natural normalization:

$$\overline{\mu_\iota} = \sqrt{\gamma_i}\sum_{i}^{-\frac{1}{2}}\mu_i,\ i \in 1, \cdots, N \tag{12}$$

where N is the number of Gaussian components of the model. Then, the divergence presented in Eq. (10) can be rewritten as the square Euclidean distance between the normalized supervector.

Finally, the derived d(·) is used for a signer model match. This is valid because of the monotony of the square function $(\cdot)^2$, we use the Euclidean distance $\|\bar{\mu}^{\alpha} - \bar{\mu}^{b}\|$ for singer model matching.

3 Dataset and Experiment

3.1 Dataset

Four datasets are included in the evaluation. Two of them are from published work—i.e., Labrosa's Artist20 [13] and Mir-1 k [17]—and the other two are newly incorporated in this work.

A new dataset **Singer107** is composed of commercial CDs, which consists of 421 Chinese pop songs by 107 singers, with 3 songs (each song lasts about 4 min) included for each singer. For classifier training and testing, we concatenate the same singer's music or recordings into a wav file. Then, the wav file is split into 20 equal parts in duration. The Python implementation of the dataset processing is published online.[1] The sample rate is equivalent to Mir-1 k and Artist20 (16 kHz and 256 kbps).

[1] https://gitlab.com/zhangxulong/project-SID-GMM-UBM.

In addition, to obtain a dataset that is closer to practical scenarios, we generate a different sample rate version dataset **Singer55** from the Singer107, which sampled at 44.1 kHz and 705 kbps. It contains 165 songs in total by 55 singers, with 3 songs from each singer. The 55 singers are selected from the first of the 107 singers with the same songs. These two datasets are available online.[2]

3.2 Experimental Setup

We use the open source RPCA implementation for the vocal separation on dataset Artist20, Singer55, and Singer107. For Mir-1 k, we only use its right channel with pure vocal. MSR Identity Toolbox [18] is used for UBM generation and GMM supervector extraction for singer identification.

We split each singer's songs into 20 equal parts. Then, 17 of them are randomly selected for training and the other 3 for testing. The following metrics are used for the evaluation of our proposed system for a different dataset in this paper: accuracy, precision, recall, and F1 value.

4 Results and Discussion

First, the optimal threshold is searched to distinguish different singer models for each dataset. Figure 2. shows the different identification results with moderate thresholds between [0,1].

In Fig. 2a, the optimal threshold value is 0.3 on Mir-1 k, and it gets the best performance of the identification. But, with the other 3 datasets, the optimal thresholds are all less than 0.1. The recognition performance also declines because the threshold is too small to make a distinction between the singers.

The threshold is directly decided by the difference between singer model likelihood value and UBM likelihood value.

The results of our singer identification experiment on different datasets is shown in Fig. 3. Artist20NoBG, Singer55NoBG, and Singer107NoBG represent the version of Artist20, Singer55, and Singer107 datasets with vocal separation respectively. It can be observed from Fig. 3 that, as the number of the singer increases, the performance of singer identification system decreases in recall and precision. A higher identification rate is obtained on Mir-1 k with clean monophonic voice. In addition, the performance on the dataset with vocal separation obviously improves.

Obviously, for the classification system, the fewer categories that it has, the higher the accuracy of the classification that it will get. However, when we compare the datasets with a different sampling rate, we find that the higher sampling rate leads to a worse identification performance.

[2]https://pan.baidu.com/s/1VJXLMmaKwIuxYrINTCizfw.

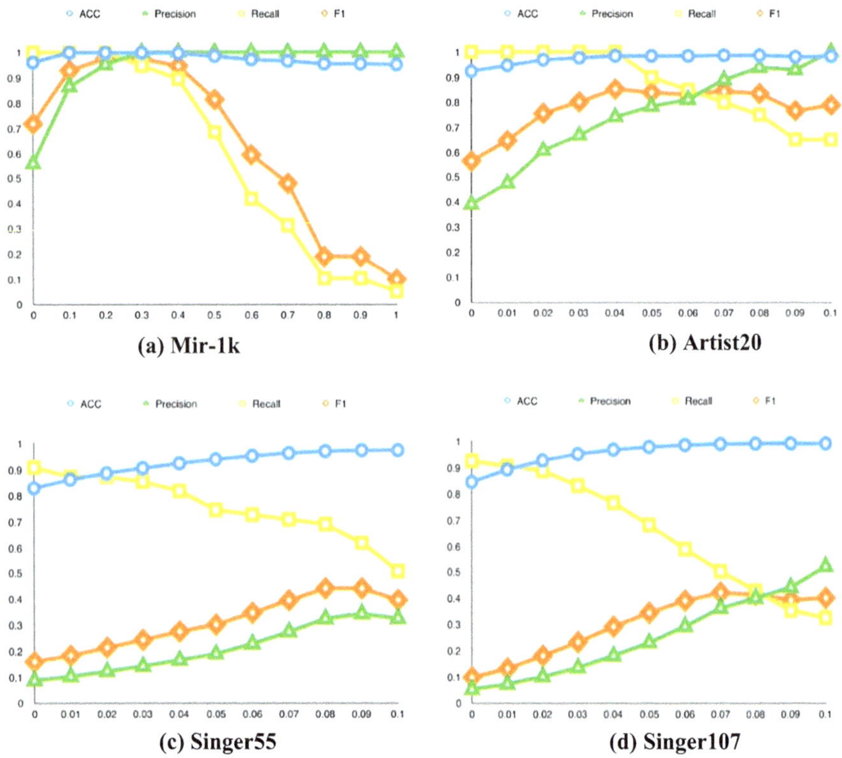

Fig. 2 Threshold value selection for classification

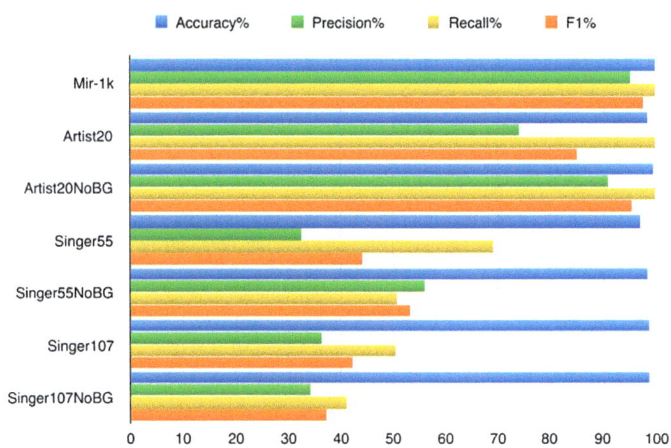

Fig. 3 Identification results on investigated datasets

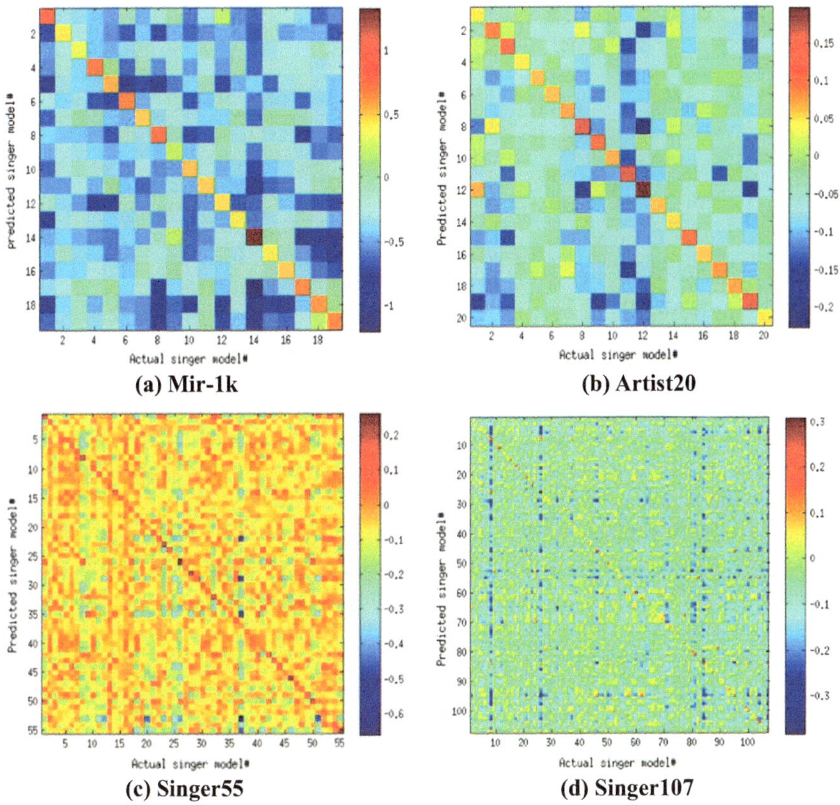

Fig. 4 Confusion matrix on different datasets

Now, we analyze the system performance on Mir-1 k, which contains only monophonic singing voice recordings. The confusion matrix of the test on Mir-1 k is given in Fig. 4a, which demonstrates relevance between the predicted singer and actual singer, and the experiment result shows that the proposed system can achieve 98% accuracy for 19 singers' recognition—the precision is 95% and the recall even achieves 100%. The experiment on Mir-1 k shows that the speaker recognition methods can be used on singer identification with the ideal state of vocal only.

The main difference between Artist20 and Mir-1 k is that the former has background music. The vocal separation operation introduces an accentuating effect for the vocal source, contributing to an obvious improvement in identification. The accuracy of Artist20 is 97.5%, and it increased 2% on Artist20NoBG, by up to 99.25%. The precision on Artist20 is 66.67%, and it increased to 94.74% on the Artist20NoBG. Improved precision is also observed on Singer55 and Singer107. However, this is accompanied by a drop of recall, indicating missed retrieval. The recall on Artist20 is 100%, and, on Artist20NoBG, it is down to 90%.

Table 1 Comparison experiment on Artist20

Method	Feature	Acc %	Pre%	Re%	F1%
Labrosa	20 MFCC	55.9	58.74	58.2	58.47
Adaboost	Echonest MFCC LPC	81	–	–	–
KNN	MFCC LPC	55	–	–	–
GMM	MFCC LPC	53	–	–	–
BPNN	MFCC Timbre	72	–	–	–
I-vector	20 MFCC	83.30	83.72	84.02	82.58
GMM-UBM	13 MFCC	99.25	94.74	90	92.31

It can be noted from Fig. 3 that the precision is improved on all investigated datasets due to vocal separation. We can conclude that the accompaniment is interference to the vocal.

The number of singers on Mir-1 k and Artist20 is just 19 and 20 respectively. We further test a large number of singers. We can also observe that, from Fig. 3, there is a decline in precision and recall in datasets Singer107 and Singer55, despite having satisfactory accuracy. From the experiments' results in Fig. 4b, c, with the singer number increasing from 20 to 55 and 107, there is a decline in precision and recall. In addition, when the sampling rate change from 16 to 44.1 k, there is a decline in precision and recall. On Singer55, the precision drops to 56% and the recall is 50.91%. On Singer107, the precision is 34.11% and the recall is 41.12%. But, the accuracy has not declined – they are 98.38% on Singer55 and 98.71% on Singer107.

Table 1 shows the identification results on Artists20, using the proposed GMM-UBM method from recently published literature [12, 19, 20] as introduced in Sect. 2. Using GMM-UBM, our proposed method outperformed all the baselines, with 16% and 10% improvement in accuracy and F1 value respectively, while i-vector is the second-best performing algorithm.

5 Conclusion

In this paper, a novel singer identification method is proposed based on GMM-UBM with vocal separation to distinguish different vocal in the music. Two new datasets are composed to complement the evaluation corpora. Our proposed method has surpassed the state-of-art algorithm on the existing dataset Artist20, with an improvement on accuracy and F1 rate of 16 and 10% respectively. The introduced vocal separation step has yielded improvement over all investigated datasets.

In the future, we will explore the impact of gender and the conversion between song and speech. In addition, we will improve the accuracy of singer identification on larger datasets.

References

1. Mesaros A (2013) Singing voice identification and lyrics transcription for music information re-trieval invited paper. In: Proceedings of IEEE 7th conference on speech technology and human-computer dialogue. Cluj-Napoca, Romania, pp 1–10
2. Hu Y, Liu G (2015) Separation of singing voice using nonnegative matrix partial co-factorization for singer identification. IEEE/ACM Trans Audio Speech Lang Process 23(4):643–653
3. Yang YH (2013) Low-rank representation of both singing voice and music accompaniment via learned dictionaries. In: Proceedings of the international symposium on music information retrieval. Curitiba, Brazil, pp 427–432
4. Sha CY, Yang YH, Lin YC, Chen HH (2013) Singing voice timbre classification of Chinese popular music. In: Proceeding of IEEE international conference on acoustics, speech and signal processing. Vancouver, Canada, pp 734–7383)
5. Huang PS, Chen SD, Smaragdis P, Hasegawa-Johnson M (2012) Singing-voice separation from monaural recordings using robust principal component analysis. In: proceeding of IEEE international conference on acoustics, speech and signal processing. Kyoto, Japan, pp 57–60
6. Ezzaidi H, Bahoura M, Rouat J (2010) Singer and music discrimination based threshold in polyphonic music. In: Proceeding of IEEE international symposium on signal processing and information technology. Luxor, Egypt, pp 445–450
7. Kim YE, Whitman B (2002) Singer identification in popular music recordings using voice coding features. In: Proceedings of the 3rd international conference on music information retrieval. Baltimore, Maryland, pp 13–17
8. Logan B (2000) Mel frequency cepstral coefficients for music modeling. In: Proceedings of the international symposium on music information retrieval. Plymouth, Massachusetts, pp 1–11
9. Nakano T, Yoshii K, Goto M (2014) Vocal timbre analysis using latent Dirichlet allocation and cross-gender vocal timbre similarity. In: Proceeding of IEEE international conference on acoustics, speech and signal processing. Florence, Italy, pp 5202–5206
10. Zhang T (2003) Automatic singer identification. In: Proceedings of international conference on multimedia and expo. Baltimore, Maryland, pp 33–36
11. Fujihara H, Kitahara T, Goto M, Komatani K, Ogata T, Okuno HG (2005) Singer identification based on accompaniment sound reduction and reliable frame selection. In: Proceedings of the international symposium on music information retrieval. London, UK, pp 329–336
12. Eghbal-Zadeh H, Lehner B, Schedl M, Widmer G (2015) i-vectors for timbre-based music similarity and music artist classification. In: Proceedings of the international symposium on music information retrieval. Malaga, Spain, pp 554–560
13. Berenzweig A, Ellis DPW, Lawrence S (2002) Using voice segments to improve artist classification of music. In: Proceeding of 22nd international conference: virtual, synthetic, and entertainment audio. Espoo, Finland, pp 1–8
14. Hasan T, Hansen JH (2011) A study on universal background model training in speaker verification. IEEE Trans Audio Speech Lang Process 19(7):1890–1899
15. Huang PS, Chen SD, Smaragdis P, Hasegawa-Johnson M (2012) Singing-voice separation from monaural recordings using robust principal component analysis. In: Proceeding of IEEE international conference on acoustics, speech and signal processing. Kyoto, Japan, pp 57–60
16. Charbuillet C, Tardieu D, Peeters G (2011) GMM-supervector for content-based music similarity. In: Proceeding of international conference on digital audio effects. Paris, France, pp 425–428
17. Hsu CL, Jang JSR (2010) On the improvement of singing voice separation for monaural recordings using the MIR-1 K dataset. IEEE Trans Audio Speech Lang Process 18(2):310–319
18. Sadjadi SO, Slaney M, Heck L (2013) MSR identity toolbox-A MATLAB toolbox for speaker recognition research. Microsoft Research, Conversational Systems Research Center, pp 1–32

19. Ratanpara T, Patel N (2015) Singer identification using perceptual features and cepstral coefficients of an audio signal from Indian video songs. EURASIP J Audio Speech Music Process 2015(1):1–12
20. Ellis DPW (2007) Classifying music audio with timbral and chroma features. In: Proceeding of the 8th international conference on music information retrieval. Vienna, Austria, pp 339–340

A Practical Singing Voice Detection System Based on GRU-RNN

Zhigao Chen, Xulong Zhang, Jin Deng, Juanjuan Li, Yiliang Jiang and Wei Li

Abstract In this paper, we present a practical three-step approach for singing voice detection based on a gated recurrent unit (GRU) recurrent neural network (RNN) and the proposed method achieves comparable results to state-of-the-art method. We combine four classic features—namely Mel-frequency Cepstral Coefficients (MFCC), Mel-filter Bank, Linear Predictive Cepstral Coefficients (LPCC), and Chroma. Then, the mixed signal is first preprocessed by singing voice separation (SVS) with the Deep U-Net Convolutional Networks. Long short-term memory (LSTM) and GRU are both proposed to solve the gradient vanish problem in RNN. In our experiments, we set the block duration as 120 ms and 720 ms respectively, and we get comparable or better results than results from state-of-the-art methods, while results on Jamendo are not as good as those from RWC-Pop.

Keywords Singing voice detection (SVD) · Gated recurrent unit (GRU) · Recurrent neural network (RNN) · Music information retrieval (MIR)

Z. Chen · X. Zhang · J. Deng · J. Li · Y. Jiang · W. Li (✉)
Department of Computer Science, Fudan University, 201203 Shanghai, China
e-mail: weili-fudan@fudan.edu.cn

Z. Chen
e-mail: zgchen15@fudan.edu.cn

X. Zhang
e-mail: xlzhang14@fudan.edu.cn

J. Deng
e-mail: jdeng17@fudan.edu.cn

J. Li
e-mail: jjli16@fudan.edu.cn

Y. Jiang
e-mail: yljiang17@fudan.edu.cn

W. Li
Shanghai Key Laboratory of Intelligent Information Processing, Fudan University, 201203 Shanghai, China

W. Li et al. (eds.), *Proceedings of the 6th Conference on Sound and Music Technology (CSMT)*, Lecture Notes in Electrical Engineering 568,
https://doi.org/10.1007/978-981-13-8707-4_2

1 Introduction

Singing voice detection (SVD) aims to localize portions of sound that containing human voice. Currently in music information retrieval (MIR), SVD is receiving increasing concerns due to its great usefulness in some singer related tasks—e.g., singer identification [1, 2], melody extraction [3].

Singing voice separation (SVS) was not frequently used as a pretreatment in SVD in many studies. However, Hennequin [4] used double stage Harmonic/Percussive Sound Separation (HPSS) [5], a simple method to separate monaural audio into harmonic and percussive components and to extract features, and got remarkable results compared with others. We choose the Deep U-Net Convolutional Networks proposed by Jansson [6], to separate singing voice directly and use the vocal signal to complete the next steps.

Traditional statistical methods with widely used speech features have been applied to SVD [7]. Following the traditional framework, they extracted a set of features and fed them to a classifier—e.g., support vector machines (SVMs) [8, 9] and random forests [10, 11]. A singing voice typically involves a higher pitch than regular speech, with wide, exaggerated intonation changes. These studies [12, 13] were often based on commonly used features in speech processing field—e.g., Mel-frequency Cepstral Coefficients (MFCC) and Linear Predictive Cepstral Coefficients (LPCC), which may not be sufficient to capture the characteristics of the accompanied singing.

It's a common phenomenon that the same features will achieve different performance in different datasets. Instead of using well-designed complex handcraft features [14], we combine four classic features—MFCC, Mel-filter Bank [4], LPCC, and Chroma [15] from MIR—and tried to depict audio characteristics more comprehensively with those features above.

GRU [16] is able to take temporal context into consideration, hence it should achieve a better performance than traditional machine learning techniques. Additionally, GRU solves the problem of gradient vanish with a simpler structure than LSTM—thus, it is preferable for real-time application. Results show that, without using temporal smoothing for post-process, our GRU-RNN can perform as well as state-of-the-art methods on public datasets.

In sum, we present a practical three-step approach for SVD in this paper. First, we separate singing voice with Deep U-Net Convolutional Networks. Second, we extract MFCC, Mel-filter Bank, LPCC, and Chroma as features. Third, we use GRU-RNN as the classifier.

This paper is organized as follows. Section 2 presents related work of this task, Sect. 3 outlines our method, Sect. 4 describes our results, and Sect. 5 presents our conclusions.

2 Related Work

SVD aims at marking out audio segments that contain human voices, which includes singing and speech, as a matter of fact. We will introduce some of the principal methods from previous studies in this section, and we'll compare our results to theirs in Sect. 4.

In an early study, Rocamora and Herrera [17] compared existing descriptors with a statistical classifier. It came out that MFCC achieved the best performance in their experiment by using their private dataset—the accuracy was 78.5%.

Ramona [9] performed this task with a large feature set, a support vector machine (SVM), and a temporal smoothing method with the Hidden Markov Model (HMM). They achieved 82% accuracy in their experiment. Mauch et al. [18] used four timbre and melody features in different combinations and fed them to support vector machine with the Hidden Markov Model (SVM-HMM) to perform the task. Their results demonstrated that the top accuracy is 87.2% when all four features were used.

Lehner et al. [10] proposed a towards light-weight and real-time online SVD system. They used only simple optimized MFCCs as the feature and the optimized random forest as the classifier. They achieved 82.36% accuracy, after manually adjusting the features and classifiers.

Observing that one of the biggest problems in automatic SVD is the confusion between vocals and instruments, based on the work of [10], Lehner et al. [11] designed a set of new audio features to reduce the amount of false vocal detections. The features consisted of fluctograms, vocal variances, spectral flatness and spectral contractions. With the new hard-craft features, results appeared to be, at least, on par with more complex state-of-the-art methods with common features.

Lehner introduced the LSTM-RNN into SVD in [14]. Different from their work in [10], which only used simple MFCCs, they combine 30 MFCCs, their delta coefficients, and other three spectral features—totaling 111 attributes. They achieved state-of-the-art results on the two publicly available databases—namely Jamendo and RWC-Pop. Furthermore, Eyben et al. [19] proposed a data-driven approach based on LSTM-RNN and standard RASTA-PLP frontend features, and results showed that LSTM-RNN outperformed all other statistical baselines. Leglaive [4] used a Bidirectional LSTM-RNN (BLSTM-RNN) as the classifier. They used just Mel-filter Bank preprocessed by HPSS, and they achieved an accuracy of 91.5% on public datasets Jamendo.

In Schlüter's work [20], pitch shifting, time stretching, and random frequency filtering were used to augment the training datasets on the public datasets Jamendo [9] and RWC-Pop [18], and the CNN model was used on Mel spectrograms to design the SVD system. Finally, the prosed method by Schlüter achieved an error rate of around 9%, which is on par with state-of-the-art results.

3 Method

We propose a three-step system. The audio signal is first preprocessed by SVS, then we extract features and fed them to the classifier. The dataset is divided into two parts, training set and testing set, and they are independent of each other. The network is first trained by the training set, then used to predict the label of the testing set. The overview of our SVD system is shown in Fig. 1. Details of every step are discussed below.

3.1 Singing Voice Separation

The audio signal is first preprocessed by SVS—we can split the mixed music signal into vocal signal and accompaniment signal through SVS. We use the Deep U-Net Convolutional Networks in [6] to accomplish this task. Datasets that we used to train the U-Nets were iKala [21] and MedleyDB [22]. The main steps are as follows.

1. Train two U-Nets respectively to predict vocal and instrumental spectrogram masks. The U-Net operates exclusively on the magnitude of audio spectrograms.
2. Compute the spectrogram mask of signal with well-trained U-Net. Apply the mask to the magnitude of original spectrum.
3. Reconstruct the signal with the new magnitude and original phase.

3.2 Feature Extraction

We describe all four features that we used in our experiment here. Many features were applied to SVD task: we picked four classic features from both speech and music field, and they were MFCC [14], Mel-filter Bank, LPCC [23], and Chroma [15].

MFCC is widely used in many speech- and audio-related tasks [17], as it characterizes the timbre of the human voice. It was found that MFCC achieved the best performance in [17]. Mel-filter Bank was extracted from a filter bank on Mel scale,

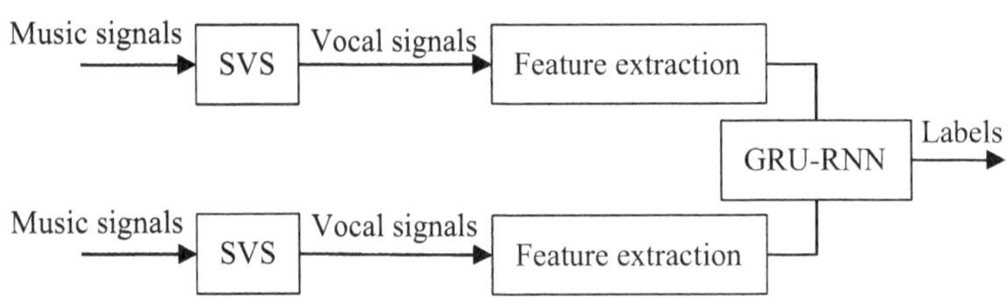

Fig. 1 A system overview

and it had a good performance in SVD [4]. LPCC had a good relationship with the vocal tract of the speaker, and it was calculated by introducing the cepstrum coefficients to the LPC parameters. Chroma, also known as Harmonic Pitch Class Profile, collects spectral energy supporting each semitone of the octave and could consider the timbre of music. It's a well-established tool for analyzing and comparing music signals [24]. We attempted to depict audio characteristics more comprehensively with the aforementioned features.

In our experiment, the frame length was set to 40 ms, with overlapping of 20 ms. On the short-scale frames stated above, we extracted 20-order MFCC, 20-order Mel-filter Bank, 12-order LPCC, and 12-order Chroma separately, then combined them to get a 64-dimensional feature.

3.3 GRU-RNN

As we know, temporal context is sometimes needful for human beings to make a vocal-nonvocal decision, and RNN could take the temporal context into consideration in classification tasks [25], so that it is appropriate to use here. The major problem of the traditional RNN is that gradient vanish, GRU, and LSTM do not solve this problem in the same way [16]. They both add the update from t to t + 1 inside these units—that's the most prominent feature that they share. In other words, both GRU and LSTM will add new content to existing content.

GRU and LSTM have a number of differences. GRU exposes its full content without any control, while LSTM controls the exposure of the memory content—i.e., GRU is simpler in structure than LSTM. Another difference is the updating of new memory content. The control of information flow in GRU is tied via the update gate, while LSTM is via the forget gate independently.

Experiments in [16] showed that convergence of GRU is often faster and the final solutions tend to be better than LSTM. Encouraged by this, we present a unidirectional RNN with a hidden layer, which consists of 60 GRU units. The input is shaped as the dimension of the combined feature and multiplied by a fixed duration block [10], and the block duration will be tuned in Sect. 4. An output layer with a single sigmoid is added. The output of the classifier is 1 or 0, 1 indicating singing and 0 indicating non–singing. Dropout is set as 0.2, and early stopping strategy is used here—it will stop if the loss of validation data gets no improvement over 5 epochs.

4 Result

In this section, we present the results on two available public datasets—i.e., RWC-Pop and Jamendo. We train several GRU-RNNs according to the previous strategy and compare the results to some methods in Sect. 2. Then, we calculate four frequently-used evaluation index [27]—i.e., frame-wise accuracy, precision, recall, and F1-measure.

4.1 Datasets

RWC Popular Music Dataset. The RWC-Pop dataset consists of 100 pop songs, with annotations that Mauch et al. [18] released. It contains 80 Japanese pop songs and 20 English pop songs. The audio files are stereo, with a sampling rate of 44.1 kHz, and the sampling precision of 16 bits. We converted the stereo files to mono wav first. The whole set is well balanced, since 51.2% of frames are singing segments and 48.8% are non-singing segments. For better comparison, we conducted the 5-fold cross validation [11]. All the data are divided into five parts: one for testing and the other four for training. The validation set is extracted 20% data from the training set. The three are independent of each other.

Jamendo Corpus. The Jamendo Corpus consists of 93 songs from Jamendo free music sharing platform with Creative Commons License. They were annotated by the same person [9]. The audio files are stereo, with a sampling rate of 44.1 kHz, Vorbis OGG format, with the sampling precision of 112 KB/s, or MP3 format, with the sampling precision of 128 KB/s. We converted the stereo files to mono wav first. The whole set is well balanced, since 50.3% of frames are singing segments and 49.7% are non-singing segments. All the data are divided into three independent parts, the same as in [9]—i.e., training set, validation set, and testing set, respectively containing 61, 16, and 16 songs.

4.2 Evaluations

For a better comprehensive evaluation, we compare predicted results of testing set to the ground truth labels and get the number of true positives (TP), false positives (FP), true negatives (TN) and false negatives (FN). Then, we calculate four frequently-used evaluation index [26]—i.e., frame-wise accuracy, precision, recall, and F1-measure.

4.3 Results and Discussion

Comparison between mixed signals and separated vocal signals. We compare the results of mixed signals and separated vocal signals in this experiment. Mixed signals are original mono signals. As previously mentioned, Deep U-Net Convolutional Networks were used to separate the vocal from the raw data. We use the combination of all the features mentioned and the parameters of GRU-RNN in Sect. 3. Block duration is set as 25 frames.

As we can see in Fig. 2, the performance in both datasets are significantly improved after the SVS. It raised the accuracy and F1-measure, by around 10% in RWC-Pop dataset and by around 15% in Jamendo dataset. So, it is useful to preprocess with

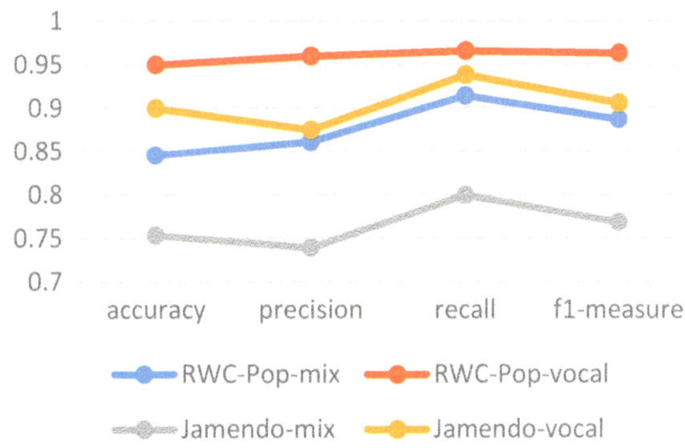

Fig. 2 Comparison between mixed signals and separated

SVS in our task here. As a matter of fact, eliminating or reducing the impacts of the accompaniment could be very helpful in various related tasks.

Fine-tuning of block duration. In previous studies, different block durations were used for the decision. Also, in our experiments, on the premise of the above parameters mentioned in Sect. 3, we found that the block duration has a great influence on our SVD system. So, we compare the results of different durations and choose an appropriate value for our experiments.

The results show that there is a positive correlation between the performance and the block duration. However, while this may seem obvious, annotation precision decreases with increased block duration. There are no black-and-white lines for annotation precision, and it is not necessarily the same in different tasks, but it is often determined by experiences.

Various durations were chosen in different studies: we choose two representative state-of-the-art methods for further comparison. Lehner [14] achieved better performance on RWC-Pop dataset, while Hennequin [4] did better with the Jamendo dataset—in fact, Hennequin [4] compared their performance with previous studies on Jamendo only. The block duration of Lehner [14] was 140 ms, and Hennequin [4] sets its duration as 800 ms. So, we finally decide to choose 120 and 720 ms to compare our results with others.

Final results. Table 1 shows our experimental results on RWC-Pop, compared with Mauch [18], Schlüter [20], Lehner-1 [11], Lehner-2 [10], Lehner-3 [14]. We set the block duration as 120 and 720 ms—they were called GRU-RNN-1 and GRU-RNN-2 respectively. Combining the results of previous studies, the best accuracy, precision, recall and F1-measure are 0.927, 0.938, 0.935, and 0.936 respectively. GRU-RNN-1 gets the best performance in recall and F1-measure and is slightly less than state-of-the-art methods on RWC-Pop. GRU-RNN-2 outperforms state-of-the-art methods on this dataset, about 3–4% above it. Given the fact that we use the combination of low-level features directly without any post-processing, our results are remarkable. Moreover, our GRU-RNN has a simpler structure than LSTM—thus, it has higher computational efficiency – so, our method is better for real-time applications than Lehner-3 [14].

Table 1 Experiment of block duration on RWC-Pop

Frames	Duration (ms)	Accuracy	Precision	Recall	F1-measure
5	**120**	**0.9205**	**0.9267**	**0.9542**	**0.9402**
7	160	0.9136	0.9080	0.9665	0.9363
11	240	0.9265	0.9271	0.9647	0.9455
17	360	0.9412	0.9514	0.9609	0.9562
25	520	0.9477	0.9560	0.9665	0.9612
35	**720**	**0.9531**	**0.9605**	**0.9696**	**0.9650**
47	960	0.9592	0.9654	0.9728	0.9691
61	1240	0.9594	0.9699	0.9671	0.9685
77	1560	0.9663	0.9726	0.9736	0.9731
95	1920	0.9617	0.9681	0.9681	0.9681
115	2320	0.9664	0.9609	0.9800	0.9704
137	2760	0.9682	0.9738	0.9644	0.9691
161	3240	0.9711	0.9765	0.9629	0.9696

Table 2 Results on RWC Popular Music Dataset

	Accuracy	Precision	Recall	F1-measure
Mauch [18]	0.872	0.887	0.921	0.904
Schlüter [20]	**0.927**	–	**0.935**	–
Lehner-1 [11]	0.875	0.875	0.926	0.900
Lehner-2 [10]	0.868	0.879	0.906	0.892
Lehner-3 [14]	0.923	**0.938**	0.934	**0.936**
GRU-RNN-1	*0.9205*	*0.9267*	***0.9542***	***0.9402***
GRU-RNN-2	***0.9531***	***0.9605***	***0.9696***	***0.9650***

Table 2 shows our experimental results on Jamendo, compared with Ramona [9], Schlüter [20], Lehner-1 [11], Lehner-2 [10], Lehner-3 [14], Leglaive [25]. We set the block duration as 120 and 720 ms, they were called GRU-RNN-1 and GRU-RNN-2 respectively.

Results on Jamendo are not as good as on RWC-Pop, as Table 3 shows. Combining the results of previous studies, the best accuracy, precision, recall and F1-measure are 0.923, 0.898, 0.926, and 0.910 respectively. GRU-RNN-1 gets the best performance in recall, which is not as good as the best results of previous studies—namely so-called state-of-the-art results on Jamendo. GRU-RNN-2 does the best in recall and F1-measure, while it is about 1% less than the best in accuracy and precision. But,

Table 3 Results on Jamendo Corpus

	Accuracy	Precision	Recall	F1-measure
Ramona [9]	0.822	–	–	0.831
Schlüter [20]	**0.923**	–	0.903	–
Lehner-1 [11]	0.882	0.880	0.862	0.871
Lehner-2 [10]	0.848	–	–	0.846
Lehner-3 [14]	0.894	**0.898**	0.906	0.902
Leglaive [4]	0.915	0.895	**0.926**	**0.910**
GRU-RNN-1	*0.8821*	*0.8539*	***0.9278***	*0.8893*
GRU-RNN-2	*0.9082*	*0.8923*	***0.9331***	***0.9122***

in general, our method gets comparable performance compared with state-of-the-art methods. There is one thing we must notice here, Schlüter [20] used pitch shifting and time stretching to make the data augmentation, while other studies were performed on original dataset. Data volume is a significant factor for machine learning, so it does not seem fair to compare our paper with Schlüter's work [20], but we still put it here for reference.

5 Conclusion

In this paper, we propose a practical three-step approach, which means good performances, simple feature selection, and higher computational efficiency for SVD based on a gated recurrent unit (GRU) recurrent neural network (RNN).

These steps are SVD, feature extraction and pattern recognition. GRU is able to take temporal context into consideration and our features are easy to extract and combine. We abandon the regular post-processing step – namely temporal smoothing.

We set the block duration as 120 and 720 ms respectively and get comparable or better performances to state-of-the-art methods with different parameters. As one can see, results on Jamendo are not as good as on RWC-Pop. Most of all, our GRU has a simpler structure and higher computational efficiency than LSTM, so it is better for real-time applications.

Future work includes the following aspects. Use the bidirectional-GRU instead of our unidirectional GRU. Find the reason why results on Jamendo are always worse than RWC-Pop and make specific improvements.

Acknowledgements This research was supported by NSFC 61671156. We thank our colleagues from Fudan University, who provided insight and expertise that greatly assisted the research, although they may not agree with all the interpretations of this paper.

References

1. Leglaive S, Hennequin R, Badeau R (2015) Singing voice detection with deep recurrent neural networks. In: Proceeding of IEEE international conference on acoustics, speech and signal processing (ICASSP). Brisbane, Australia, pp 121–125
2. Kim YE, Whitman B (2002) Singer identification in popular music recordings using voice coding features. In: Proceedings of the 3rd international conference on music information retrieval. Paris, France, pp 13–17
3. Salamon J, Gómez E (2012) Melody extraction from polyphonic music signals using pitch contour characteristics. IEEE Trans Audio Speech Lang Process 20(6):1759–1770
4. Leglaive S, Hennequin R, Badeau R (2015) Singing voice detection with deep recurrent neural networks. In: Proceeding of IEEE international conference on acoustics, speech and signal processing (ICASSP). Brisbane, Australia, pp 121–125
5. Ono N, Miyamoto K, Le Roux J, Kameoka H, Sagayama S (2008) Separation of a monaural audio signal into harmonic/percussive components by complementary diffusion on spectrogram. In: Proceeding of 16th European signal processing conference. Lausanne, Switzerland
6. Jansson A, Humphrey E, Montecchio N, Bittner R, Kumar A, Weyde T (2017) Singing voice separation with deep U-Net convolutional networks. In: Proceeding of 18th international society for music information retrieval conference. Suzhou, China
7. Sonnleitner R, Niedermayer B, Widmer G, Schlüter J (2012) A simple and effective spectral feature for speech detection in mixed audio signals. In: Proceedings of the 15th international conference on digital audio effects (DAFx'12). York, UK
8. Vembu S, Baumann S (2005) Separation of vocals from polyphonic audio recordings. In: Proceeding of international society for music information retrieval conference, London, UK, pp 337–344
9. Ramona M, Richard G, David B (2008) Vocal detection in music with support vector machines. In: Proceeding of IEEE international conference on acoustics, speech and signal processing (ICASSP). Las Vegas, USA, pp 1885–1888
10. Lehner B, Sonnleitner R, Widmer G (2013) Towards light-weight, real-time-capable singing voice detection. In: Proceeding of international society for music information retrieval conference. Curitiba, Brazil, pp 53–58
11. Lehner B, Widmer G, Sonnleitner, R (2014) On the reduction of false positives in singing voice detection. In: Proceeding of IEEE international conference on acoustics, speech and signal processing. Florence, Italy, pp 7480–7484
12. Regnier L, Peeters G (2009) Singing voice detection in music tracks using direct voice vibrato detection. In: Proceeding of IEEE international conference on acoustics, speech and signal processing. Taipei, Taiwan, pp 1685–1688
13. Pikrakis A, Kopsinis Y, Kroher N, Díaz-Báñez JM (2016) Unsupervised singing voice detection using dictionary learning. In: Proceeding of 24th European signal processing conference. Budapest, Hungary, pp 1212–1216
14. Lehner B, Widmer G, Bock S (2015) A low-latency, real-time-capable singing voice detection method with LSTM recurrent neural networks. In Proceeding of 23rd European signal processing conference. Nice, France, pp 21–25
15. Ellis DPW, Poliner GE (2007) Identifying cover songs' with chroma features and dynamic programming beat tracking. In: Proceeding of IEEE international conference on acoustics, speech and signal processing. Honolulu, USA, pp 1429–1432

16. Chung J, Gulcehre C, Cho K, Bengio Y (2014) Empirical evaluation of gated recurrent neural networks on sequence modeling. arXiv preprint, 1412.3555
17. Rocamora M, Herrera P (2017) Comparing audio descriptors for singing voice detection in music audio files. In: 11th Brazilian symposium on computer music. São Paulo, Brazil, pp 27–36
18. Mauch M, Fujihara H, Yoshii K, Goto M (2011) Timbre and melody features for the recognition of vocal activity and instrumental solos in polyphonic music. In: Proceeding of international society for music information retrieval conference. Miami, Florida, pp 233–238
19. Eyben F, Weninger F, Squartini S, Schuller B (2013) Real-life voice activity detection with LSTM recurrent neural networks and an application to hollywood movies. In: Proceeding of IEEE international conference on acoustics, speech and signal processing. Vancouver, Canada, pp 483–487
20. Schlüter J, Grill T (2015) Exploring data augmentation for improved singing voice detection with neural networks. In: Proceeding of international society for music information retrieval conference. Malaga, Spain, pp 121–126
21. Chan TS, Yeh TC, Fan ZC, Chen HW, Su L, Yang YH, Jang R (2015) Vocal activity informed singing voice separation with the iKala dataset. In: Proceeding of 2015 IEEE international conference on acoustics, speech and signal processing. Brisbane, Australia, pp 718–722
22. Bittner RM, Salamon J, Tierney M, Mauch M, Cannam C, Bello JP (2014) MedleyDB: a multitrack dataset for annotation-intensive MIR research. In: Proceeding of international society for music information retrieval conference, vol 14. Taipei, Taiwan, pp 155–160
23. Gupta H, Gupta D (2016) LPC and LPCC method of feature extraction in speech recognition system. In: Proceeding of 6th international conference cloud system and big data engineering. Noida, India, pp 498–502
24. Muller M, Ewert S, Kreuzer S (2009) Making chroma features more robust to timbre changes. In: Proceeding of IEEE international conference on acoustics, speech and signal processing. Taipei, Taiwan, pp 1877–1880
25. Leglaive S, Hennequin R, Badeau R (2015) Singing voice detection with deep recurrent neural networks. In: Proceeding of IEEE international conference on acoustics, speech and signal processing (ICASSP). Brisbane, Australia, pp 121–125
26. Sokolova M, Lapalme G (2009) A systematic analysis of performance measures for classification tasks. Inf Process Manage 45(4):427–437

Multimodel Music Emotion Recognition Using Unsupervised Deep Neural Networks

Jianchao Zhou, Xiaoou Chen and Deshun Yang

Abstract In most studies on multimodal music emotion recognition, different modalities are generally combined in a simple way and used for supervised training. The improvement of the experiment results illustrates the correlations between different modalities. However, few studies focus on modeling the relationships between different modal data. In this paper, we propose to model the relationships between different modalities (i.e., lyric and audio data) by deep learning methods in multimodal music emotion recognition. Several deep networks are first applied to perform unsupervised feature learning over multiple modalities. We, then, design a series of music emotion recognition experiments to evaluate the learned features. The experiment results show that the deep networks perform well on unsupervised feature learning for multimodal data and can model the relationships effectively. In addition, we demonstrate a unimodal enhancement experiment, where better features for one modality (e.g., lyric) can be learned by the proposed deep network, if the other modality (e.g., audio) is also present at unsupervised feature learning time.

Keywords Music emotion recognition · Multimodal learning · Deep neural networks

1 Introduction

Music plays an important role in human history, especially in the digital age [1]. Since music itself is an expression of emotions, it is natural for us to search and

J. Zhou · X. Chen (✉) · D. Yang
Institute of Computer Science and Technology, Peking University, 128 Zhongguancun North Street, Haidian District, Beijing 100080, China
e-mail: Xiaoou.Chen@pku.edu

J. Zhou
e-mail: Jianchou.Zhou@pku.edu

D. Yang
e-mail: Deshun.Yang@pku.edu

W. Li et al. (eds.), *Proceedings of the 6th Conference on Sound and Music Technology (CSMT)*, Lecture Notes in Electrical Engineering 568,
https://doi.org/10.1007/978-981-13-8707-4_3

organize music by emotional contents [2]. Music emotion recognition has gained increasing attention over the past decade.

In the music emotion recognition field, audio data has been commonly used and audio-based techniques could obtain promising results [3, 4]. As a complementary source, lyric data contains rich semantic information of songs and more emotionally relevant information that is not included in audio data [5]. Many experiments have proved that audio and lyric data correlate with each other for music emotion recognition. Experiments about multimodal music emotion recognition, in which audio and lyric data are both applied, have achieved promising results [6–8]. Experiments in [6, 7] showed that audio and lyric data were complementary, and the performance was further improved when combining both of the two modalities. Multiple modalities, including audio and lyric data, were employed using different fusion methods in [8, 9]. The performance of the experiments had a significant improvement compared with the previous researches.

Most studies on multimodal music emotion recognition mainly pay attention to the improvement of recognition results, in which different modalities (i.e., lyric and audio) are just simply combined and used for supervised training. Few studies focus on modeling the relationships between lyric and audio data.

Deep learning approaches that are based on multimodal data have gained much attention in recent years. Srivastava and Salakhutdinov [10] proposed a Deep Boltzmann Machine (DBN) model to extract a meaningful representation of multimodal data, including image and text. Experiment results showed that the multimodal DBM could learn a good generative model of the joint space of multimodal inputs and the learned features performed well on discriminative tasks. Ngiam et al. [11] presented a series of tasks based on novel deep networks for multimodal learning. Their models were validated on audio-visual speech classification tasks, demonstrating promising performance and effective shared representation learning. Liu et al. [12] adopted a deep learning approach to construct effective models with a multimodal dataset (EEG signals, eye movement features, and peripheral physiological signals) to recognize emotions, and the proposed models could extract high-level representation features that were effective for emotion recognition. Many experiments have shown that deep learning models can achieve a good performance on a multimodal feature learning task.

In this paper, we propose to model the relationships between different modalities (i.e., lyric and audio data) via deep learning method in music emotion recognition. For a single song, lyric data and audio data in this study refer to the feature vectors extracted from the lyric file and audio file of the song. Since music emotion is represented in the PAD model [13] in this paper, we will focus on music emotion regression task, in which emotion space is viewed as continuous. Inspired by the good performance of deep learning models on multimodal feature learning task [11, 12], we employ the deep networks in [11] to model the relationships between lyric and audio data. Concretely, several deep networks are first applied to perform unsupervised feature learning over multiple modalities. We, then, design a series of music emotion regression experiments to evaluate the learned features. Through the performance of these features in the music emotion regression task, we can infer whether the deep networks are effective in modeling the relationships between different modalities.

We will consider the learning settings shown in Fig. 1. The overall task can be divided into three phases: unsupervised feature learning, supervised regression model training, and regression testing. Several deep networks are applied to unsupervised feature learning. The SMOreg, an algorithm of support vector machine regression, is used for supervised regression model training and testing to evaluate the different representations learned by deep networks. Specifically, three experiments, including unimodal learning, multimodal learning and unimodal enhancement, are taken into consideration.

In the unimodal learning experiment, only one modality is available at all three phases and a representation is learned for every single modality by the deep network. We use the emotion recognition results of this experiment as baseline results. In the multimodal learning experiment, data from all modalities are used at all three phases. We employ a deep network to model the relationships between audio and lyric data. In the unimodal enhancement experiment, data from multiple modalities is available during unsupervised feature learning, while only data from single modality is provided during supervised regression model training and testing. Another deep network is used in this experiment and the aim is to explore whether a better single modality representation can be learned when unlabeled data from multiple modalities is given.

The remainder of this paper is organized as follow. Section 2 introduces the building blocks of the mentioned deep networks, Sect. 3 presents the architecture of the mentioned deep networks, Sect. 4 details the database construction and feature extraction, Sect. 5 shows the results and analysis, and Sect. 6 concludes the paper.

	Unsupervised feature learning	Regression model training	Testing
Unimodal learning	Audio	Audio	Audio
	Lyric	Lyric	Lyric
Multimodal learning	A+L	A+L	A+L
Unimodal enhancement	A+L	Audio	Audio
	A+L	Lyric	Lyric

Fig. 1 The overall experiment settings

2 Building Blocks

Unsupervised models like Auto Encoder (AE) or Restricted Boltzmann Machine (RBM) can be used to learn a new representation of the original input. As building blocks of the multimodal deep networks, they will be described in detail in the following content.

2.1 *Auto Encoder*

AE consists of an encoder part and a decoder part [14]. The encoder transforms an n-dimensional vector x into a d-dimensional hidden representation y through a deterministic mapping:

$$y = f_{\theta}(x) = s(Wx + b) \tag{1}$$

s is a nonlinear function, such as the sigmoid function. The resulting hidden representation y is, then, mapped back by the decoder to a reconstructed n-dimensional vector z:

$$z = g_{\theta'}(y) = s\left(W'y + b'\right) \tag{2}$$

The aim is to make x and z similar as much as possible. Parameters are obtained by minimizing the average reconstruction error L(x; z). L(x; z) has many types, such as the reconstruction cross-entropy:

$$L(x, z) = -\sum_{k=1}^{n}\left[x_k \log z_k + (1 - x_k)\log(1 - z_k)\right] \tag{3}$$

2.2 *Restricted Boltzmann Machine*

The RBM is an undirected graph model, which has one visible layer and one hidden layer, with only interlayer connections. The standard type of RBM, named Bernoulli RBM, has binary-valued hidden and visible units and consists of a weight matrix $W = \left\{w_{ij}\right\}$ associated with the connection between hidden unit h_j and visible unit v_i, as well as bias terms a_j for the hidden units and b_i for the visible units [15]. Given this, the energy function is defined as:

$$E(v, h; \theta) = -\sum_i \sum_j w_{ij} v_{ij} h_j - \sum_i b_i v_i - \sum_j a_j h_j \tag{4}$$

The joint distribution over the visible units and hidden units could be calculated with energy function as:

$$P(v, h; \theta) = \frac{\exp(E(v, h; \theta))}{Z(\theta)} \tag{5}$$

$Z(\theta)$ is the normalization constant. Given a set of visible variables $\{v_m\}_{m=1}^{M}$, the derivative of log-likelihood with respect to weight can be obtained by:

$$\frac{1}{M}\sum_{i=1}^{M}\frac{\partial \log P(v_m; \theta)}{\partial w_{ij}} = E_{P_{data}}\left[v_i h_j\right] - E_{P_{model}}\left[v_i h_j\right] \tag{6}$$

where E represents mathematical expectation. RBM can be trained by various algorithms, such as Contrastive Divergence (CD) algorithm [12]. Also, in this paper, Bernoulli RBM is used.

3 Deep Networks

In this section, we describe the deep networks for the task of unsupervised feature learning, where the inputs of audio data and lyric data are both feature vectors with real values. In the unimodal learning experiment, two baseline feature learning models are trained for audio and lyric data respectively. In the multimodal learning experiment where both modalities are used at all phases, we adopt a Bimodal Deep Auto Encoder (BDAE) model to learn a shared representation of audio and lyric data. When both modalities are present during feature learning but only one modality is available for supervised regression model training and testing, the Unimodal Deep Auto Encoder (UDAE) model is applied to learn an enhanced representation of a single modality. Detailed descriptions are as follow.

3.1 Unimodal Learning

One of the most straightforward approaches is training a feature learning model for every single modality. After training the model, the calculated hidden variables are used as a new representation for the input data. The learned new representation can, then, be used for supervised regression model training and testing. Experiments show that the hidden representation learned by AE performs better in emotion recognition than that learned by RBM for audio data, while lyric data is exactly opposed to audio data. So, AE is used to train audio data and RBM is used to train lyric data at unsupervised feature learning stage in this experiment. As shown in Fig. 2, we use these two models denoted by audio-AE and lyric-RBM to perform unsupervised

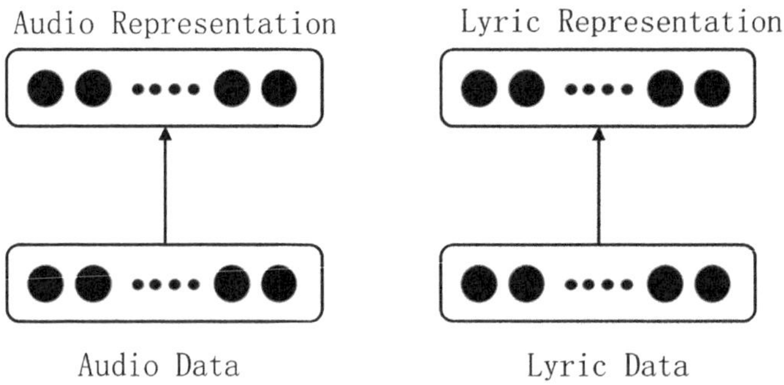

Fig. 2 Audio-AE and lyric-RBM models

feature learning for a single modality. The learned representations are trained to build regression models and the recognition results of regression models are regarded as our baseline results.

3.2 Multimodal Learning

In the multimodal learning experiment, both modalities are present at all three phases. The BDAE model shown in Fig. 3 is used to model the relationships between audio and lyric data. To train the BDAE model, we first train audio-AE model for audio data and lyric-RBM model for lyric data. Two hidden layers, indicated by h_L and h_A, are then concatenated directly. The joint hidden layer is treated as the input layer of an upper AE model. When unfolding them into a BDAE, we keep the related weights tied. From Fig. 3, we can see that W_1 and W_1^T have tied weights, W_2 and W_2^T have tied weights, and W_3 and W_3^T have tied weights.

At last, the unsupervised back-propagation algorithm is applied to finetune the weights. The learned shared representation is, then, used for supervised regression model training and testing. From the emotion recognition results of the corresponding regression model, we can infer whether the BDAE network models the relationships effectively.

3.3 Unimodal Enhancement

When both modalities are used during unsupervised feature learning but only a single modality is present for supervised regression model training and testing, the UDAE models are trained to reconstruct both modalities with only one modality at unsupervised feature learning stage and, thus, discover correlations across the two modalities. The lyric-UDAE model shown in Fig. 4, which only use lyric data

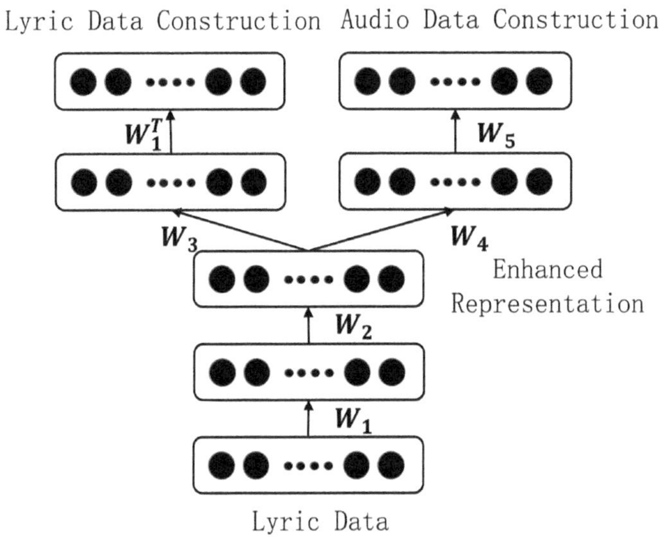

Fig. 3 BDAE model

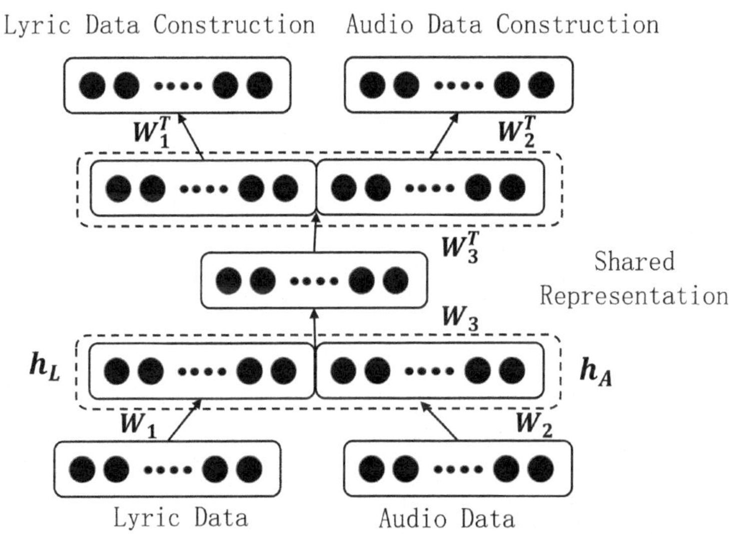

Fig. 4 Audio-AE and lyric-RBM models

to reconstruct two modalities, is first initialized by the previous trained lyric-RBM model. The hidden layer of the lyric-RBM is, then, treated as the visible layer of the upper AE model. When unfolding them to reconstruct input data, the weights W_1 and W_1^T are tied, while other weights could be initialized randomly. Unsupervised back-propagation is needed to finetune the parameters. Finally, the learned enhanced representation can be used for supervised regression model training and testing. From the music emotion recognition results of the corresponding regression model, we can know if the lyric-UDAE network learns a better representation for lyric data when audio data is given. Correspondingly, there is an audio-UDAE model, which only uses audio data to reconstruct two modalities—the training method of which is similar to that of lyric-UDAE model, except that the audio-UDAE model is initialized by the pre-trained audio-AE model.

4 Experiment Settings

4.1 Dataset

Our initial dataset contains more than 3000 Chinese songs that were downloaded from music websites, including audio and lyric files for each song, which cover more than 900 singers and more than 1000 albums, and different genres, such as pop, rap, and rock. The songs are annotated with PAD values ranging from -4 to 4 by 11 volunteers, whose age from 22 to 50 by using Self-Assessment Manikins [16]. When a song is annotated by more than 8 volunteers and the emotion values given by different annotators are consistent, meaning that all positive or all negative, the song will get a mean value as its emotion label and be retained in our final dataset. This way, exactly 2000 songs are preserved to make up our final dataset.

4.2 Input Feature Vectors for Deep Networks

Audio Feature Vector. We first extract features from wav format files of the dataset by jAudio [17], which is a system to extract basic features from an audio signal. The signal sampling rate is 22 kHz and the used window size is 512 ms. The commonly used 27 features are extracted to compose a feature vector of 112 dimensions for a song [18].

Since MIDI features are often used in music emotion recognition field, we also apply them in this work. We first transform wav format files of the dataset to MID format files by WIDI. Then, the MIDI features of a single song can be extracted from the transformed MID format file by jSymbolic, which is a feature extraction system for extracting high-level musical features from symbolic music representations, such as the MID format files [19]. The resulted 1022-dimensional feature vector of a single song is composed of 102 kinds of MIDI features [18].

The two feature vectors are concatenated directly to obtain the final audio feature vector. Normalization is performed on it, all dimensions of this are a real value, ranging from 0 to 1. The normalized audio feature vector is used as the audio modality input for the deep networks.

Lyric Feature Vector. The downloaded lyric files of all songs are first processed with traditional NLP tools, consisting of stop-words filtering and word segmentation etc. N-Gram features are, then, extracted from the processed lyrics. N-Gram refers to a distinctive term containing N consecutive words appeared in documents. In our work, Unigram (N = 1), Bigram (N = 2), and Trigram (N = 3) features are extracted from the lyrics. To reduce the lyric feature space, we select the 2000 most frequently appeared N-grams (n = 1, 2, 3) as lyric features [18]. So, the lyric feature vector can be expressed as $(v_1, v_2, \ldots v_{2000})$. Here $v_i \in \{0, 1\}$: if N-Gram appeared in the lyric, $v_i = 1$; otherwise $v_i = 0$.

4.3 *Related Parameters of Deep Networks*

The hidden units of audio-AE and lyric-RBM models are both 200. For two UDAE models, the dimensions of the first hidden layer and the shared representation layer are both 200. For the BDAE model shown in Fig. 4, the dimension of the concatenated hidden layer is 400 and the shared representation units are 200. Learning rate for all models is 0.1 and training epochs ranges between 100 and 300.

4.4 *Evaluation Criteria*

We conduct a series of experiments to evaluate the performance of different representations learned by deep networks. Specifically, the representations learned by deep networks are trained to build regression models with SMOreg algorithm for each emotion dimension (P, A, and D), and the performance of regression models is measured in terms of the correlation coefficient (CF), which is defined as follows:

$$CF_{XY} = \frac{\sum_{i=1}^{N}\left(R(X_i) - \overline{R(X_I)}\right)\left(Y_i - \overline{Y}\right)}{\sqrt{\sum_{i=1}^{N}\left(R(X_i) - \overline{R(X_I)}\right)^2}\sqrt{\sum_{i=1}^{N}\left(Y_i - \overline{Y}\right)^2}} \quad (7)$$

where Y_i is the emotion label, R(X_i) is the regression value of feature vector X_i. More clearly, experiment results are the correlation coefficients between the scores assigned by various regression models and the mean score assigned by volunteers on each dimension. The high correlation coefficient value illustrates the high effectiveness of the learned representation. We use 5-fold cross-validation method to evaluate the performance of regression models and the final mean CF value is shown.

5 Result Analysis

5.1 *Unimodal Learning*

In the unimodal learning experiment, audio data is used to train audio-AE model, while lyric data is used to train lyric-RBM model. As our baseline results, the performance of two regression models based on the representation of audio-AE and the representation of lyric-RBM are shown in Table 1.

Of the two modalities, we can conclude that lyric data has a greater contribution to dimension P than audio data, while audio data contributes more to dimension A and D than the lyric. The reason for this is that audio data contains energy-related information, which reflects emotion intensity more directly, and content-related information, which reflects music concepts more directly. In contrast, lyric data includes more semantic information, which indicates emotional inclinations more directly.

5.2 *Multimodal Learning*

In the multimodal learning experiment, both modalities are available during all three stages. We investigate if a meaningful shared representation of the two modalities can be learned. Thus, we can infer whether the BDAE network is effective in modeling the relationships between audio and lyric data. Experiment results are shown in Table 2.

Table 1 Performance of the regression models trained on the representations of baseline Audio-AE and Lyric-RBM networks

Feature learning model	P	A	D
Audio-AE	0.574	**0.824**	**0.602**
Lyric-RBM	**0.756**	0.295	0.415

The first column indicates which network's representation the regression model is trained on

Table 2 Comparison among the regression models trained on the representations of two baseline networks and the shared representation of BDAE network

Feature learning model	P	A	D
Audio-AE	0.574	**0.824**	0.602
Lyric-RBM	0.756	0.295	0.415
BDAE	**0.799**	0.815	**0.677**

The first column indicates which network's representation the regression model is trained on

As aforementioned in Sect. 5.1, the regression model based on the representation of audio-AE performs well in dimension A and D, but badly in dimension P, while the regression model based on the representation of lyric-RBM performs exactly the opposite. When the shared representation from BDAE model is used, the performance improves and CF values of all emotion dimensions are relatively large.

This indicates that the two modalities provide useful and complementary information for music emotion regression and greater improvement can be achieved when both two modalities are used. The BDAE network is proved to be effective in modeling relationships and can learn a meaningful shared representation from two modalities. However, it is worth noting that the regression model based on the representation of BDAE network performs worse in dimension A. The reason may be that audio data is highly discriminative for dimension A, involving a lyric in a way as simple as concatenation could hurt the performance.

Since the goal of our experiments is not to improve the accuracy of music emotion recognition, but to explore whether our unsupervised deep networks can model the relationships between different modalities effectively. We do not compare the results of our experiments with those of other studies in this paper.

5.3 Unimodal Enhancement

In the unimodal enhancement experiment, we evaluate if we can learn a better representation for one modality when given multiple modalities during unsupervised feature learning. As shown in Table 3, the regression model based on the enhanced representation of lyric-UDAE shows a significant improvement compared with the baseline model and performs better in dimension P and A, while slightly worse in dimension D. The lyric-UDAE network can learn a better representation for lyric data through unimodal enhancement learning, in which the audio data corresponding to the lyric is also used as reconstruction targets. The energy and dynamics related information, which reflects emotion intensity, and symbolic information, which reflects music concepts directly, contained in audio data, could contribute to the enhanced representation and, then, to the performance of lyric data. Of course, the slightly worse performance in dimension D indicates that the requirement of audio data reconstruction in lyric representation learning restrains the representation in some way.

Table 3 Comparison between the regression models trained on the representation of baseline Lyric-RBM network and the enhanced representation of Lyric-UDAE network

Feature learning model	P	A	D
Lyric-RBM	0.756	0.295	**0.415**
Lyric-UDAE	**0.775**	**0.356**	0.411

The first column indicates which network's representation the regression model is trained on

It is worth noting that the regression model based on the representation of audio-UDAE do not improve recognition results compared with the baseline regression model based on the representation of audio-AE. Audio data are highly discriminative for emotion regression, and the requirement of lyric reconstruction in representation learning restrains the audio representation. Simply adding lyric data information can sometimes hurt the performance of audio data.

6 Conclusion

In this paper, we are interested in modeling the relationships by deep learning models between different modalities (i.e., lyric and audio) in the music emotion recognition field. In our experiment, several deep networks are first applied to perform unsupervised feature learning over multiple modalities. We, then, design a series of music emotion regression experiments to evaluate the learned features. Experiment results show that audio and lyric data for music emotion regression have correlations at a "mid-level" and the unsupervised deep learning method can model the relationships effectively. Deep networks can learn related information from the two modalities by their strong unsupervised feature learning ability. Meanwhile, a better representation can be learned for one modality by the proposed deep network if the other modality is used as supervision information in feature learning.

Our future work is focused on two aspects. Firstly, more deep network structures are to be explored to model better relationships between the two modalities. Secondly, we need to propose a feasible solution for the practical issue that all modalities are available in the phase of model training, while some modalities of a music track are not available when the emotion of the track is to be predicted.

References

1. Yang YH, Lin YC, Su YF, Chen HH (2008) A regression approach to music emotion recognition. IEEE Trans Audio Speech Lang Process 16(2):448–457
2. Kim YE, Schmidt EM, Migneco R, Morton BG, Richardson P, Scott J, Speck JA, Turnbull D (2010) Music emotion recognition: a state of the art review. ResearchGate 86(00):937–952
3. Laurier C, Grivolla J, Herrera P (2008) Multimodal music mood classification using audio and lyrics. In: International conference on machine learning and applications, pp 688–693
4. Tzanetakis G, Ermolinskyi A, Cook P (2003) Pitch histograms in audio and symbolic music information retrieval. J New Music Res 32(2):143–152
5. Hu X, Downie JS, Ehmann AF (2009) Lyric text mining in music mood classification. In: International society for music information retrieval conference, ISMIR 2009, pp 411–416. Kobe International Conference Center, Kobe, Japan, October
6. Yang YH, Lin YC, Cheng HT, Liao I, Ho YC, Chen HH (2008) Toward multi-modal music emotion classification. In: Pacific Rim conference on multimedia: advances in multimedia information processing, pp 70–79, (2008)

7. Hu X, Downie JS (2010) Improving mood classification in music digital libraries by combining lyrics and audio. In: Joint international conference on digital libraries, JCDL 2010, pp 159–168, Gold Coast, Queensland, Australia, June
8. Lu Q, Chen X, Yang D, Wang J (2010) Boosting for multi-modal music emotion classification. In International society for music information retrieval conference, ISMIR 2010, pp 105–110, Utrecht, Netherlands, August
9. Zhao Y, Yang D, Chen X (2010) Multi-modal music mood classification using co-training. In: International conference on computational intelligence and software engineering, pp 1–4
10. Srivastava N, Salakhutdinov R (2012) Multimodal learning with deep boltzmann machines. J Mach Learn Res 15(8):1967–2006
11. Ngiam J, Khosla A, Kim M, Nam J, Lee H, Ng AY (2011) Multimodal deep learning. In: International conference on machine learning, ICML 2011, pp 689–696. Bellevue, Washington, USA, June 28–July
12. Liu W, Zheng WL, Lu BL (2016) Emotion recognition using multimodal deep learning
13. Mehrabian A (1995) Framework for a comprehensive description and measurement of emotional states. Genet Soc Gen Psychol Monogr 121(3):339
14. Zhou J, Peng L, Chen X, Yang D (2016) Robust sound event classification by using denoising autoencoder. In: 18th IEEE international workshop on multimedia signal processing, MMSP 2016, pp 1–6. Montreal, QC, Canada, September 21–23
15. Chen H, Murray AF (2003) Continuous restricted boltzmann machine with an implementable training algorithm. Vis Image Signal Process IEE Proc 150(3):153–158
16. Lang PJ (1980) Behavioral treatment and bio-behavioural assessment: computer applications. Technology in mental health care delivery systems. Norwood Ablex
17. Mckay C, Fujinaga I, Depalle P (2005) jAudio: a feature extraction library
18. Guan D, Chen X, Yang D (2012) Music emotion regression based on multi-modal features. In: Proceedings of international symposium on computer music modeling and retrieval, pp 70–77
19. Mckay C, Fujinaga I (2006) Symbolic: a feature extractor for midi files, pp 302–305

Music Summary Detection with State Space Embedding and Recurrence Plot

Yongwei Gao, Yichun Shen, Xulong Zhang, Shuai Yu and Wei Li

Abstract Automatic music summary detection is a task that identifies the most representative part of a song, facilitating users to retrieve the desired songs. In this paper, we propose a novel method based on state space embedding and recurrence plot. Firstly, an extended audio feature with state space embedding is extracted to construct a similarity matrix. Compared with the raw audio features, this extended feature is more robust against noise. Then recurrence plot based on global strategy is adopted to detect similar segment pairs within a song. Finally, we proposed to extract the most repeated part as a summary by selecting and merging the stripes containing the lowest distance in the similarity matrix under the constraints of slope and duration. Experimental results show that the performance of the proposed algorithm is more powerful than the other two competitive baseline methods.

Keywords Music summary detection · Extended audio feature · State space embedding · Recurrence plot · Global strategy

---

Y. Gao · Y. Shen · X. Zhang · S. Yu · W. Li (✉)
School of Computer Science and Technology, Fudan University, Shanghai 201203, China
e-mail: weili-fudan@fudan.edu.cn

Y. Gao
e-mail: 16110240016@fudan.edu.cn

Y. Shen
e-mail: 16110240016@fudan.edu.cn

X. Zhang
e-mail: xlzhang14@fudan.edu.cn

S. Yu
e-mail: 18110240029@fudan.edu.cn

W. Li
Shanghai Key Laboratory of Intelligent Information Processing, Fudan University, Shanghai 201203, China

W. Li et al. (eds.), *Proceedings of the 6th Conference on Sound and Music Technology (CSMT)*, Lecture Notes in Electrical Engineering 568,
https://doi.org/10.1007/978-981-13-8707-4_4

1 Introduction

Music is a temporal sequence with a length of several minutes. It is difficult for listeners to quickly capture the most attractive parts in popular songs [1]. The listeners can use fast forward or rewind operation to identify the most attractive parts in popular songs, but it is time-consuming once they met a new song. In this condition, automatically summarizing the whole song using the most attractive part is timely helpful.

Music summary just refers to the most representative and attractive part of a song. A song is a highly structured signal and generally includes five parts: introduction, verse, chorus, bridge, and ending [2]. *The chorus contains the main idea, or big picture, of what is being expressed lyrically and musically. It is repeated throughout the song, and the melody and lyric rarely vary.*[1] Hence the chorus sections become the good candidates for music summary. Automatic music summary detection is a task that recognizes the most representative part, normally the repeated part of a song. And it is also a very important technique applied to ringtone producing, quick music browsing, music retrieval boost, and music data storage improvement etc.

At present, the previous algorithms can be divided into two categories: similarity matrix-based algorithms and machine learning-based algorithms. Similarity matrix-based algorithms [3–7] usually include the following steps. Firstly, audio features such as MFCC and Chroma are extracted. Similarity matrix (SM) is then built on extracted features. Time-lag diagram is further generated to represent the similar snippets within the processed song. Secondly, shape detection algorithms applied for image processing detect similar snippets in SM. Finally, the most repeated part will be extracted by clustering and grouping these similar segment pairs in the light of music theory. The other algorithms are based on machine learning [8–11]. Music signal is first divided into several segments, then those segments are processed by classifying or clustering, which produces the final representative part.

Each category has its own merits and drawbacks. Similarity matrix-based algorithms can achieve satisfactory accuracy and computing speed, while they are always subject to weak robustness of audio features. Further, due to the significant distinction between SM and natural images, these methods are limited by directly using shape detection algorithms applied for image processing to detect the similar segment pairs in SM. The machine learning based methods, to a certain extent, alleviate the influence of weakly robust features. However, owing to the uncertainty of results with machine learning algorithms and dependency on datasets, these methods always subject to training dataset, training environment and training times. In addition, many algorithms based on the latter are always processing under the condition that tempo in a song is constant.

Since there exists uncertainty and too many constraints with the machine learning based methods, we propose a method based on self-SM. Instead of the raw audio features adopted in [3–5], an extended audio feature with state space embedding is extracted to compute self-SM, which is robustness to noise and irrelevant to the key

[1] http://en.wikipedia.org/wiki/Song_structure.

shift. Then we introduce the recurrence plot to binaries SM to extract possibly similar pairs. Finally, the most repeated part regarded as a summary of the processed song will be extracted by clustering and grouping these similar pairs.

This paper is organized as following: self-SM irrelevant to the key shift is calculated in Sect. 2, where we introduce state space embedding to enhance the robustness of raw audio feature. Similar subsegment pairs are detected with recurrent plot and the most representative snippet is selected by two principles we proposed in Sect. 3. In Sect. 4, the experiment for the detection of similar subsegment pairs and the experiment of music summary detection are described. Finally, some conclusions and future work are suggested in Sect. 5.

2 Self-similarity Matrix Irrelevant to a Key Shift

To detect the most repeated part of a song, we first construct a self-SM with audio features where every two audio frames belonging to the same song are compared.

2.1 Feature Extraction

In general, the repeated part refers to the snippets with the same melody. Note that they are usually played on different instruments in different keys and tempos. So it is hard for a computer to identify the repetition of a melody. Chroma [12], a frequency-based feature, has been widely used in representing music signal due to its ability to depict musical melody and harmony.

In this paper, we first extract an improved Chroma features by using the chroma DCT-reduced log pitch (CRP) toolbox [13] with Hamming windows of 200 ms and 50% overlap. The frequency information of each frame is quantified into 12 frequency bins each of which is irrelevant to the octave. Compared with original chroma features, it is more robust against the timbre difference. Given a CRP feature $C_{i,t}$ generated above, which is a 12-dimensional vector, the result of feature extraction for each song is shown as:

$$C = \{C_{i,t} | i = 1, 2, \ldots, 12;\; t = 1, 2, \ldots, T\} \tag{1}$$

where t and T are the index and the total number of frames respectively and i is the index of each element in the 12-dim feature vector, $1 \leq i \leq 12$.

2.2 Extended and Centralized Chroma Feature

Though the chroma feature is robust against the timbre difference, it is still easily subject to various noise. To enhance its robustness, we introduce a time-delay embedded feature representation with state space embedding which is widely used in the nonlinear series analysis [14] and has recently been applied in many Music Information Retrieval (MIR) tasks [15–17].

This feature representation is to extend the feature vector of a single frame by concatenating the feature vectors of its neighboring frames. In this way, it is able to acquire more temporal information to resist the noise contained in the single frame and remove slight tempo-invariance in audio signals. As shown in Fig. 1, raw feature matrix C is extended into improved feature matrix E with state space embedding.

While state space embedding could enhance the robustness of audio features, it generally blurs the exact positions of starting points and end points of the similar pairs detected in the following sections. To alleviate the negative effect, a Kaiser window [18] is applied, where the middle elements corresponding to the current frame is larger than the edge elements corresponding to the neighboring frames. Hence the neighboring frames are weakened but the current frame is enhanced.

For an audio signal, the center-weighted extended feature E is calculated as follows:

$$\begin{aligned} D_i(t) &= [C_{i,t-n}, \ldots, C_{i,t}, \ldots, C_{i,t+n}] \otimes w \\ E(t) &= [D_1(t), D_2(t), \ldots, D_{12}(t)] \end{aligned} \tag{2}$$

where t is the index of the current frame, n is the bidirectional expansion factor, w is a Kaiser window. This equation means that the tth frame is centralized and weighted by concatenating its left n and right n neighbors. After this process, the center-weighted extended feature $E(t)$ is expanded to $(12 \times (2 \times n + 1))$ for the tth frame.

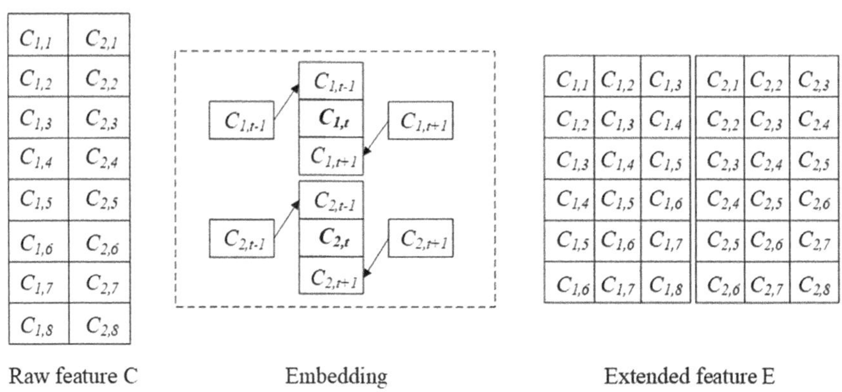

Fig. 1 Time-delay extended feature representation with state space embedding

2.3 Extended and Centralized Chroma Feature

Musical key in a song is not generally constant and key shift commonly occurred in chorus snippets of popular songs. The snippets with a similar melody have a difference of one or more semitones. Typically, the chorus part always rises a semitone to enhance the climax of the song. Through analysis, a key shift can just be represented with the extended and centralized chroma feature E by cyclic shift. Equation (3) means that the nth frame is i semitones higher than the mth frame.

$$E(m) \approx shift_i(E(n)) \tag{3}$$

where $shift_i(x)$ means cyclically shifting down the feature x by $i \times (2 \times n + 1)$. After shift processing for the nth frame, the feature obtained is consistent with that of the mth frame.

Self-SM denotes the similarity between every two frames in a time sequence. In order to enhance its robustness to key shift, the self-SM in this paper is defined as:

$$\begin{aligned} D_{m,n} &= \arg \min_{1 \le i \le 12} \varphi(E(m), shift_i(E(n))) \\ S(m,n) &= 1/D_{m,n} \end{aligned} \tag{4}$$

where $S(m, n)$ represents the similarity between the mth frame and the nth frame. And $\varphi(x, y)$ denotes the Euclidean distance between x and y.

3 Music Summary Detection

Self-SM has been constructed with the extended and centralized feature. Next, we extract the most repeated snippets by detecting all possibly similar subsegment pairs on self-SM.

3.1 Recurrence Plot Based on Global Threshold Policy

In this section, the self-SM is first binarized to obtain more precise time positions of similar stripes. The binary version is called recurrence plot where the value 1 in the point (i, j) means that the ith frame of one music signal is similar to the jth frame of the same music signal based on a specific threshold policy, and the value 0 indicates that the two frames are dissimilar. Here we adopt a globally sparse strategy on the self-SM to generate the recurrence plot.

The globally sparse strategy has been successfully used for cover song detection and chord recognition [16, 17]. The value in the self-SM is quantified as 1 if it belongs not only to the top $\eta\%$ minimum points in its row but also to that in its column, otherwise, it is quantified as 0. For given the self-SM S, the recurrence plot based on globally sparse strategy is defined as:

$$M(i,j) = \begin{cases} 1 & S(i,j) \in row_\eta(i) \cap col_\eta(j) \\ 0 & otherwise \end{cases} \tag{5}$$

where the $row_\eta(i)$ and $col_\eta(j)$ refer to the top η minimum points of the ith row and the jth column respectively in S.

3.2 Similar Subsegment Pairs Detection

In fact, not all white stripes in a recurrent plot obtained above represent similar subsegment pairs. In order to select the right similar pairs, we propose two principles.

As shown in Fig. 2, two points with value 1 (white point in the plotted figure) in plot matrix M means that they belong to a same similar stripe when they are adjacent to each other, horizontally, vertically, or in the secondary diagonal direction. For example, the point a and its right neighbor point b belong to the same stripe.

Inspired by [19], those points in the similarity matrix with the lowest distance value usually indicate true matching. Hence we further select those stripes in M holding the points with the lowest distance value in S. Given S and M as input, the detailed procedure of detecting N possibly similar segment pairs is described as below.

Firstly, the stripe from M that holds the point with the lowest value (highest similarity) in S is picked out, and the positions of its starting and end point are recorded. Next, those points belonging to the extracted stripe in S and M are removed to prevent duplicate extraction. Finally, repeat the first step and the second step until N stripes are extracted or no more stripes can be extracted. To refine the results, we further add some constraints to the above process:

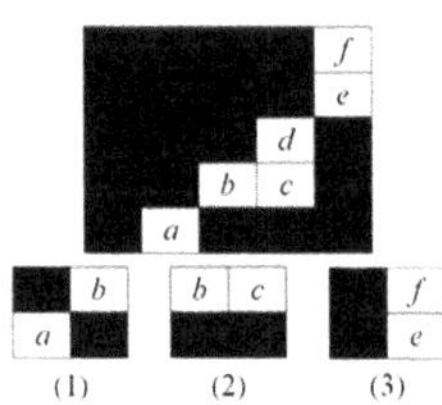

Fig. 2 Three cases where two white points in M are regarded as belonging to one stripe: (1) the secondary diagonal direction, (2) horizontal direction, (3) vertical direction

(a) The duration t_{min} of each detected subsegment should be longer than 8 s since we observed that the duration of the summary snippet usually ranges from 8 to 50 s in our experiments.
(b) The slope α of each detected stripe should range from 0.5 to 2, i.e. the difference in tempo between similar pairs is less than 2 times. When α is equal to 1, it means that two segments corresponding to this stripe have the same tempo. In fact, due to the state space embedding and recurrent plot effectively enhanced the robustness of the matrix M, this constraint has little work in our experiment. Note that slope α of a stripe is calculated only based on its start point and end point.

After the above processing, N truly similar stripes are selected in matrix M. As shown in Fig. 3, we display the selection result of Chinese song called *AiRuChaoShui* where the possibly similar stripes are clear and sparse. If a stripe starts at point A (x_1, y_1) and ends at point B (x_2, y_2), it means that the snippet $[x_1, x_2]$ is similar with the snippet $[y_1, y_2]$. The summary part of this song labeled by human ranges from the 143rd to 180th frame and the selected stripes by our method almost fall within this range.

Now all possibly similar segment pairs have been selected. In order to select the most repeated snippet, another principle is proposed in this section. Firstly, if the snippet n_1 is similar to the snippet n_2, they are put into the same set. Then if the snippet n_3 is similar with the snippet n_4, and the overlap between n_1 and n_3 accounts for more than 75% of the average length of two snippets, all sets including the four snippets are merged. Repeat the first step and the second step until no more sets can be extended. Finally, pick out the largest set and extract the highest overlap part as the summary snippet of this processing song.

Fig. 3 A sample of similar segments extraction

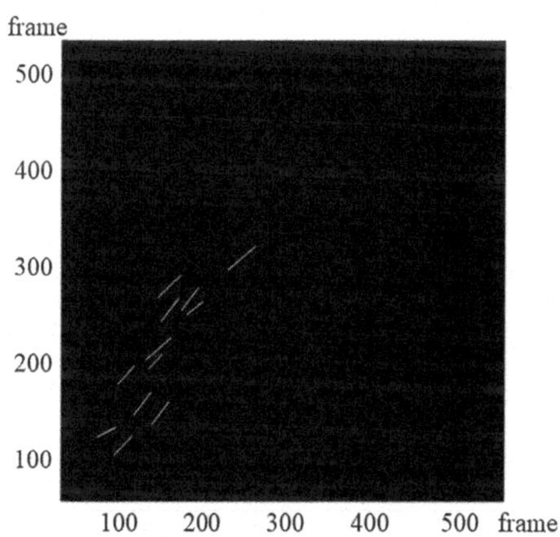

4 Experiments

4.1 Experimental Datasets

The experimental dataset comprises of 240 popular songs. They are of different ages, singers and genres. In light of language category: English, Chinese and other, these songs are divided into 3 data sets each of which includes 80 songs. And the correct summary of each song is labeled by a human.

For example, Chinese dataset contains not only rock like *NvFangDeShengMing*, but also love songs like *WenBie*. And for the song called *LiuXingYu*, its labeled summary includes four snippets: [1:10–1:35], [2:24–2:48], [3:22–3:46], [3:47–4:15]. After summary detection algorithm we proposed, the detected result is [3:17–3:15]. Note that [1:10] refers to the time point of 1 min and 10 s.

4.2 Experiments for Similar Subsegment Pairs Detection

The effectiveness of similar subsegment pairs detection is first evaluated. Given that M refers to the number of correctly similar pairs detected and N the total number of possibly similar pairs detected, M/N is regarded as the precision of detection. Note that in this section the similar subsegments include not only the summary snippets labeled above but also all snippets with a similar melody. And the similar pairs are considered correct if the correctly similar subsegment accounts for more than 80% of total length.

Detection results based on three datasets are shown in Table 1. Obviously, around 20 similar subsegment pairs are detected within a song. In Chinese, English and other datasets, the possibly similar pairs detected are, on average, 19.5, 20.4 and 21.1 respectively. And detection accuracy is larger than 70%. We also observe that the results are stable in different datasets, hence it is almost irrelevant to the language category. Since the incorrectly similar pairs account for a smaller proportion, this step is considered effective for the following summary detection.

Table 1 Experimental results for similar subsegment pairs detection

	Chinese	English	Other
N	19.5	20.4	21.1
M/N	0.73	0.72	0.70

Table 2 Experimental results for music summary detection

	P_p	P_r	F
Matthew and Jonathan [20]	0.64	0.52	0.54
Bartsch and Wakefield [4]	0.71	0.82	0.72
Our method	**0.73**	**0.85**	**0.83**

4.3 *Experiments for Music Summary Detection*

In this section, three metrics precision, recall and F measure are adopted to evaluate the method we proposed. They are defined as follows:

$$P_p = \frac{x \cap z_i}{|x|}, \qquad i = argmax_j |x \cap z_j|$$
$$P_r = \frac{x \cap z_i}{|z_i|}, \qquad i = argmax_j |x \cap z_j|$$
$$F = \frac{2 \times P_p \times P_r}{P_p + P_r} \tag{6}$$

where x and $\{z_j\}$ refer to the summary snippet detected by our method and all summary snippets labelled by human respectively. Snippet z_i has the largest overlap with x. We define that the proportion of summary snippet detected incorrect summary snippet as precision, which is represented by P_p. And the proportion of correct summary snippet labelled by a human in the summary snippet we detect is defined as recall, which is denoted by P_r.

Our proposed method for summary detection is compared with Matthew and Jonathan [20] and Bartsch and Wakefield [4] on the 240 popular songs. The results are shown in Table 2. Obviously, the performance of our method outperforms the two baseline methods. The precision, recall and F measure reach 73, 85 and 83% respectively.

5 Conclusions

In this paper, we proposed a novel music summary detection algorithm with state space embedding and recurrence plot. To enhance its robustness, we first introduce a time-delay embedded feature representation with state space embedding. Self-SM is then constructed based on the embedded feature. Recurrence plot is used to binarize the self-similarity matrix to extract the possibly similar subsegment pairs. Finally, the most repeated part, as a summary of the processed song, will be extracted by the two principles we proposed. State space embedding and recurrent plot could be applied to all time series. Hence there is potentially room to introduce more features to work for summary detection. In the future work, we will try to adopt multiple features to improve the performance of the detection algorithm.

References

1. Gao S, Li H (2015) Popular song summarization using chorus section detection from audio signal. In: Proceedings of the 17th international workshop on multimedia signal processing (MMSP), pp 1–6. IEEE, Xiamen, China
2. Maddage NC, Xu C, Kankanhalli MS et al (2004) Content-based music structure analysis with applications to music semantics understanding. In: Proceedings of the 12th ACM international conference on multimedia (MM), pp 112–119. ACM, New York, USA
3. Matthew C, Jonathan F (2002) Automatic music summarization via similarity analysis. In: Proceedings of the 3rd international society for music information retrieval (ISMIR), pp 122–127. Paris, France
4. Bartsch MA, Wakefield GH (2005) Audio thumbnailing of popular music using chroma-based representations. IEEE Trans Multimedia (MM) 7(1):96–104
5. Lu L, Zhang HJ (2003) Automated extraction of music snippets. In: Proceedings of the 11th ACM international conference on multimedia (MM), pp 140–147. ACM, CA, USA
6. Chai W (2006) Semantic segmentation and summarization of music: methods based on tonality and recurrent structure. IEEE Signal Process Mag 23(2):124–132
7. Nieto O, Humphrey EJ, Bello JP (2012) Compressing music recordings into audio summaries. In: Proceedings of 13th international society for music information retrieval (ISMIR), pp 313–318, Porto, Portugal (2012)
8. Xu C, Maddage MC, Shao X (2005) Automatic music classification and summarization. IEEE Trans Speech Audio Process (TASLP) 13(3):441–450
9. Xu C, Zhu Y, Tian Q (20025) Automatic music summarization based on temporal, spectral and cepstral features. In: Proceedings of international conference on multimedia and expo, pp 117–120, Lausanne, Switzerland
10. Zlatintsi A, Maragos P, Potamianos A (2012) A saliency-based approach to audio event detection and summarization. In: Proceedings of the 20th European signal processing conference (EUSIPCO), pp 1294–1298, Bucharest, Romania
11. Logan B, Chu S (2000) Music summarization using key phrases. In: Proceedings of the IEEE international conference on acoustics, speech, and signal processing (ICASSP), pp 749–752. Istanbul, Turkey
12. Müller M, Ewert S (2010) Towards timbre-invariant audio features for harmony-based music. IEEE Trans Audio Speech Lang Process (TASLP) 18(3):649–662
13. Müller M, Ewert S (2011) Chroma Toolbox: MATLAB implementations for extracting variants of chroma-based audio features. In: Proceedings of the 12th international conference on music information retrieval (ISMIR), pp 215–220, Miami, Florida
14. Kantz H, Schreiber T (2004) Nonlinear time series analysis. Cambridge University Press, Cambridge, United Kingdom
15. Bello JP (2011) Measuring structural similarity in music. IEEE Trans Audio Speech Lang Process (TASLP) 19(7):2013–2025
16. Serrà J, Serra X, Andrzejak RG (2009) Cross recurrence quantification for cover song identification. New J Phys 11(9):093017
17. Cho T, Bello JP (2011) A feature smoothing method for chord recognition using recurrence plots. In: Proceedings of the 12th international society for music information retrieval (ISMIR), pp 651–656, Miami, Florida

18. Bertin-Mahieux T, Ellis DPW (2011) Large-scale cover song recognition using hashed chroma landmarks. In: Proceedings of IEEE workshop on applications of signal processing to audio and acoustics (WASPAA), pp 117–120, New York, USA
19. Egorov A, Linetsky G (2008) Cover song identification with IF-F0 pitch class profiles. MIREX extended abstract
20. Matthew C, Jonathan F (2003) Summarizing popular music via structural similarity analysis. In: Proceedings of IEEE workshop on applications of signal processing to audio and acoustics (WASPAA), pp 1159–1170, New York, USA (2003)

Constructing a Multimedia Chinese Musical Instrument Database

Xiaojing Liang, Zijin Li, Jingyu Liu, Wei Li, Jiaxing Zhu and Baoqiang Han

Abstract Throughout history, more than 2000 Chinese musical instruments have existed or been historically recorded, they are of non-negligible importance in Chinese musicology. However, the public knows little about them. In this work, we present a multimedia database of Chinese musical instruments. This database includes, for each instrument, text descriptions, images, audio clips of playing techniques, music clips, videos of the craft process and recording process, and acoustic analysis materials. Motivation and selecting criteria of the database are introduced in detail. Potential applications based on this database are discussed, and we take the research on subjective auditory attributes of Chinese musical instruments as an example.

Keywords Chinese musical instruments · Multimedia database · Subjective auditory attributes · Audio recording dataset

1 Introduction

In this section, we first introduce the background of Chinese musical instruments. Next, related works on musical instruments are briefly summarized. Finally, we elaborate on the motivation of constructing the database of Chinese musical instruments.

X. Liang (✉) · Z. Li · J. Zhu · B. Han
China Conservatory of Music, Beijing 100101, China
e-mail: lxj123456789886@gmail.com

J. Liu
Communication University of China, Beijing 100024, China

W. Li
Fudan University, Shanghai 201203, China

W. Li et al. (eds.), *Proceedings of the 6th Conference on Sound and Music Technology (CSMT)*, Lecture Notes in Electrical Engineering 568,
https://doi.org/10.1007/978-981-13-8707-4_5

1.1 Background

Chinese musical instruments are the physical embodiment of Chinese traditional music culture and an important part of the splendid Chinese traditional culture. The acoustic characteristic of a musical instrument directly reflects the cultural background and aesthetic accumulation of different historical periods and nationalities. Establishing a multimedia database of Chinese musical instruments is of great significance: it not only provides authoritative and reliable fundamental data support for the creation, performance, theory, education, popularization and research of Chinese music, but is also a platform for people around the world to understand Chinese traditional musical instruments from various perspectives. Such a database should collect audio, video, text, and images of Chinese musical instruments, so that people can have a full image of a specific musical instrument.

1.2 Related Works

Research on musical instrument database first appeared in museology. It focuses on digitalizing the sound collections of musical instruments. From 2009 to 2011, MIMO (Musical Instrument Museums Online) [1] consortium created the world's largest freely accessible database of musical instruments. In China, an instrument database called "China National Music Instrument Digital Museum" was jointly established by the Central Conservatory of Music, Huazhong University of Science and Technology and Huazhong Normal University [2]. Another type of musical instrument database has emerged in the area of music information retrieval (MIR). Such a database usually includes diverse annotations, which are used for various machine learning based MIR research tasks, such as melody extraction, rhythm analysis, instrument recognition, music affect calculation, etc. [3]. A typical example is the Real World Computing (RWC) music database created in Japan, which includes audio clips of 50 musical instruments from Japan and the west [4]. References [5–8] are some other music datasets that include musical instruments that were created for MIR. In these datasets, western musical instruments occupy the greatest proportion—most of the audio clips are only a few seconds long, which is often inadequate for researchers.

1.3 Motivation

Research into Chinese musical instruments occupy an indispensable position in Chinese music research: however, a comprehensive and high-quality Chinese instrument database has not yet appeared. Therefore, a unified standard and highly professional database is of great necessity—the establishment of such a database is of great significance to the in-depth study of Chinese music culture.

In recent years, the authors have been working on the construction of a Chinese musical instrument database to meet multiple research needs, such as musical instruments acoustics and performance analysis. Previous projects include "Standard Library of Chinese Musical Instruments" and "Acoustic and Spectrum Analysis of Chinese Musical Instruments" [9, 10]. By carefully selecting recording environment, recording devices, audio samples, and players, these works provide abundant data for sound quality evaluation, performance study, and subjective experiment of musical instruments.

2 Description of the Database

Apart from the instruments from the same origin that resemble each other in both sound and appearance, there are almost 800 different instruments in China. In our work, 200 musical instruments are chosen with regional representativeness. In this part, we detail the construction and characteristics of multimedia Chinese music instruments database (http://47.90.12.82).

2.1 *Database Construction*

To construct a high-quality database, especially one with high-quality sound, criteria for every aspect have to be carefully considered. In this part, we give special emphasis on selection considerations of music instruments, recording environment, devices, players, and recording contents.

Currently, the database includes Chinese orchestral instruments, such as Erhu, Pipa, Ruan, and Chinese opera accompaniment instruments, such as Banhu and Jinghu. Instruments of ethnic minorities and reformed instruments will be included in the future.

To exhibit the instrument sounds and the delicate playing skills that represent the characteristics of Chinese traditional musical instruments, we perform recording and acoustic measurements in a fully anechoic room. As seen in Fig. 1, B&K free field microphones, installed on 3/4 spherical stent with a diameter of 3 m, are set up to record Ruan and Zheng.

The sound of an instrument is a result of playing behaviour. Therefore, the players who are recorded should be proficient in musical playing, with standardized skills acknowledged by authorities, instead of personalized skills, such that typical sounds of the corresponding musical instrument can be generated. In our work, the players are professional players with high-level skills. Moreover, the physical and psychological status of the player should be in normal condition.

Recording contents include an open string (for string instruments) or lowest tone (for wind instruments) for different dynamics (including forte, mezzo forte, and piano), scales, playing techniques, and music clips. During the acoustic analysis

Fig. 1 Recording environment of Ruan and Zheng

process, the open string or lowest tone is used to determine the stability of intonation. Playing on different dynamics embodies consideration of the influence of dynamics on timbre. Scales are used to determine the range of the instrument and playing techniques provide samples for acoustic analysis of the effect on the sound blend.

The database can be searched both by text and category. Instruments are categorized by Hornbostel-Sachs taxonomy (the most widely accepted classifying system for musical instruments by ethnomusicologists and organologists), materials, the way of performance, and period. Users can choose one category under each of them and the system will return an intersection of all selected categories.

2.2 *Characteristics of Multimedia Chinese Musical Instruments*

2.2.1 Completeness

First, musical instruments included in the database cover all basic types of the Hornbostel-Sachs taxonomy—i.e., chordophones, idiophones, membranophones, and aerophones. With regards to the content, all recorded sounds are carefully labelled in detail, including pitch, playing technique, record location, players, etc. Tables 1 and 2 take "Guqin" as an example, illustrating the basic information and file labels.

2.2.2 Multi-functionality

The multimedia database of Chinese musical instruments provides comprehensive information about each instrument. It contains not only the basic information (Hornbostel-Sachs taxonomy, materials, the way of performance, period, and text introduction), but also high-fidelity audio recordings, instrument images from dif-

Table 1 Basic information of Guqin

Name	Guqin
1st class H-S	Chordophone
2nd class H-S	Friction chordophone
Material	Silk
Performance	Plucked
Place	China
Period	Ancient, modern, contemporary

Table 2 Example of file information of Guqin

File name	Chenkunli style Guqin-A1A
File type	Audio
Format	Wave
Player	Sui Y/Zhang M
Location	NIMAC
RecDate	2013.09.23
Track	4
MicDir	F, B, U, D
UplDate	2017.10.17

ferent angles, and videos of the craft process and the recording process. Figure 2 are examples of picture and video screenshot of Pipa.

In addition, the database provides acoustic analysis of the instruments, including range, pitch deviation (in terms of equal temperaments), harmonic analysis, spectrum analysis, sonority analysis, temporal analysis of harmonics, etc. An example of spectrum analysis in the decay stage of Pipa is illustrated in Fig. 3.

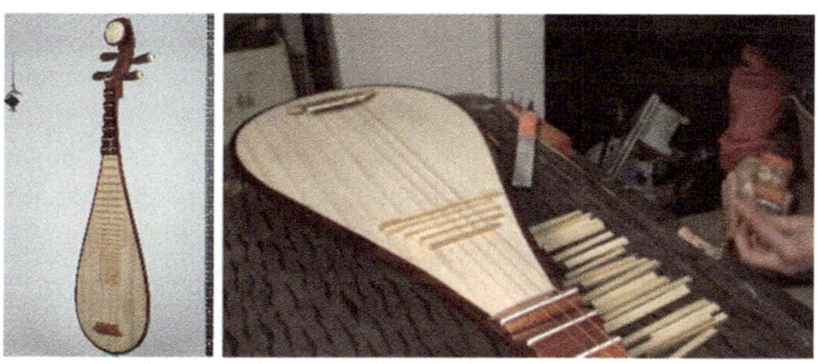

Fig. 2 Picture and screenshot of Pipa

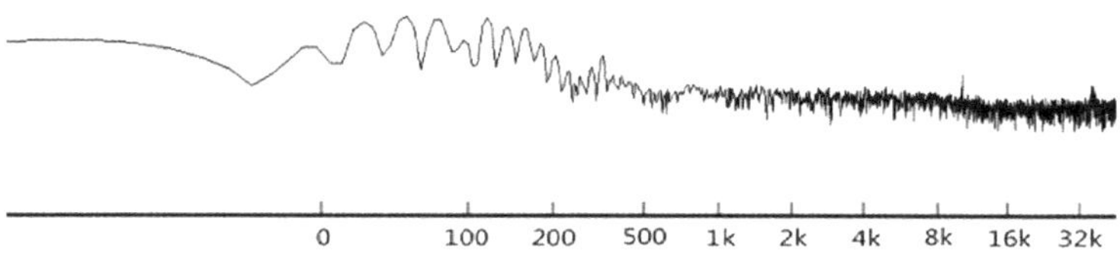

Fig. 3 Spectrum analysis in decay stage of A2 of Pipa

2.3 Extensibility

At present, recording work of 70 Chinese traditional musical instruments has been finished and more than 700 files have been uploaded. Thanks to the above specifications of database construction, the multimedia database of Chinese musical instrument can be consistently and continuously enriched in the near future.

3 Potential Applications

Multimedia Chinese musical instrument database provides abundant data support for studies of the Chinese musical instruments, music acoustics, instrumentation and orchestration of Chinese orchestra, and music informatics.

Below, we take "Sharpness Rating" as an example to introduce subjective auditory attribute research based on the database. The study chooses 37 Chinese musical instruments that are most commonly used in Chinese orchestra, then it chooses a 4s-long test music clips for each instrument. Before the experiment, the average sound pressure of the audio clips of different instruments are calibrated, and the sound pressure during the experiment is controlled within the range of 64–70 dB. Thirty-four participants with more than 10 years of music experience participate in the experiment. Participants are first asked to listen twice to all audio clips of the 37 musical instruments in a row to get an overall impression. Then, they are asked to listen for a third time to rate the extent of "sharpness" of each instrument, ranging from 1 to 9. Method of scale rating is used to analyze the result, as shown in Fig. 4. Jinghu, the musical instrument that is rated as the sharpest, is a representative musical instrument in Peking Opera. Bass Gehu, invented by Yusen Yang in 1950s, and is considered to be the least sharp musical instrument among the participants. Further research on subjective auditory attributes of different registers of a certain musical instrument will provide a reference for improvement of musical instrument design.

Apart from the applications mentioned above, the database helps researchers to analyze playing techniques used in a piece of music, which can be further applied to performance study and computer-aided instrumental performance education. One can also compare acoustic characteristics of Chinese musical instruments and that of theirs to analyze similarities in timbre across different culture. By studying the spectrum of different stages of a certain instrument, it can also be helpful for the

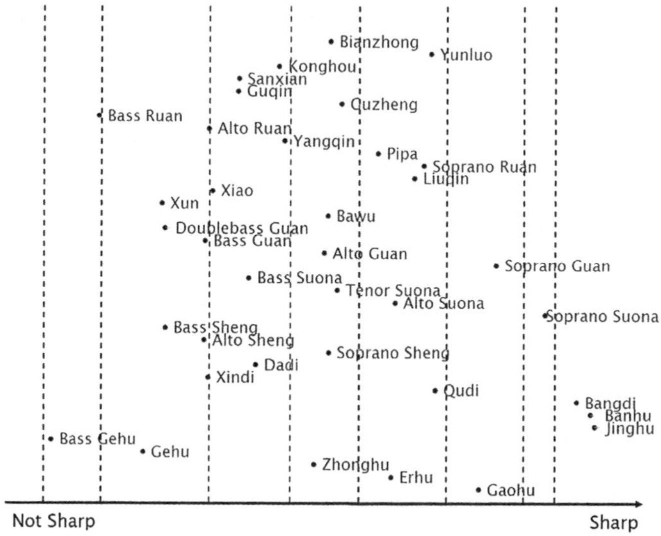

Fig. 4 "Sharpness rating" of 37 Chinese musical instruments

sound synthesis of Chinese music instruments. In addition, composers from around the world can have easy access to basic information for composing pieces with Chinese musical instruments.

4 Conclusion

In this paper, we present a multimedia Chinese musical instrument database. Contents and characteristics of the database are introduced and considerations of musical instruments selection, recording environment, devices, players, and recording contents are discussed. An example of an application involving "Sharpness Rating" of 37 common Chinese musical instruments is introduced and other potential applications of the database are discussed.

In future work, we will further extend the database to cover more Chinese musical instruments. We also plan to formulate the standard of musical instruments of ethnic minorities and reformed instruments and to provide more data support for all related research communities.

Acknowledgements This research is funded by the Ministry of Culture of PRC, the Ministry of Science and Technology, Beijing Municipal Social Science Foundation (17YTC028). We are also thankful to all participants who participated in the work through the years.

References

1. The MIMO (Musical Instruments Museums Online) project. (2017-10-02) [2018-08-30]. http://www.mimo-international.com/MIMO/
2. China National Music Instrument Digital Museum project. (2015-05-05) [2018-08-30]. http://www.cmimo.org/about
3. Li W, Li Z, Gao T (2018) Understanding digital music—a review of music information retrieval technology. In the invited talk of the fifth China Conference on Sound and Music Technology (CSMT). J Fudan Univ (Nat Sci Edn) 57(3), 350–392
4. Goto M, Hashiguchi H, Nishimura T et al (2003) RWC music database: music genre database and musical instrument sound database. In: Proceeding of ISMIR, pp 229–230
5. Peeters G (2004) A large set of audio features for sound description (similarity and classification) in the CUIDADO project
6. Livshin A, Rodet X (2004) Musical instrument identification in continuous recordings. In: Digital audio effects, pp 1–1. Naples, Italy
7. Donnelly P, Sheppard J (2015) Cross-dataset validation of feature sets in musical instrument classification. In: IEEE international conference on data mining workshop, pp 94–101
8. Shabtai N, Behler G, Vorländer M, Weinzierl S (2017) Generation and analysis of an acoustic radiation pattern database for forty-one musical instruments. J Acoust Soc Am, 1246–1256
9. Han B (2009) Is there an acoustics standard for music instruments? The psychophysics, basic of China national musical instruments acoustics standard library. Entertainment Technol 2:46–48
10. Fu X (2009) Significance of establishing criteria for the sound system of Chinese folk musical instruments. J Xinjiang Normal Univ (Soc Sci) 30(4):110–112

Acoustic Sound Processing and Analysis

Bird Sound Detection Based on Binarized Convolutional Neural Networks

Jianan Song and Shengchen Li

Abstract Bird Sound Detection (BSD) is helpful for monitoring biodiversity and in this regard, deep learning networks have shown good performance in BSD in recent years. However, such a complex network structure requires high memory resources and computing power at great cost for performing the extensive calculations required, which make it difficult to implement the hardware in BSD. Therefore, we designed an audio classification method for BSD using a Binarized Convolutional Neural Network (BCNN). The convolutional layers and fully connected layers of the original Convolutional Neural Network were binarized to two values. The Area Under ROC Curve (AUC) score of BCNN achieved comparable results with the CNN in an unseen evaluation. This paper proposes two networks (CNNs and BCNNs) for the BSD task of the IEEE AASP Challenge on the Detection and Classification of Acoustic Scenes and Events (DCASE2018). The Area Under ROC Curve (AUC) score of BCNN achieved comparable results with CNN on the unseen evaluation data. More importantly, the use of the BCNN could reduce the memory requirement and the hardware loss unit, which are of great significance to the hardware implementation of a bird sound detection system.

Keywords Bird sound detection · Convolutional neural networks · Binarized neural network

1 Introduction

Many biological monitoring projects rely on the acoustic detection of birdsong. Bird sound detection (BSD) involves identifying the presence of bird sounds in a given audio recording [1]. In many wildlife projects involving conventional remote moni-

J. Song (✉) · S. Li
Beijing University of Posts and Telecommunications, Beijing 100876, China
e-mail: songjianan@bupt.edu.cn

S. Li
e-mail: shengchen.li@bupt.edu.cn

W. Li et al. (eds.), *Proceedings of the 6th Conference on Sound and Music Technology (CSMT)*, Lecture Notes in Electrical Engineering 568,
https://doi.org/10.1007/978-981-13-8707-4_6

toring, the monitoring/detection process is not fully automated and requires intensive manual labor to label the obtained data [2, 3]. Consequently the automatic detection of birdsong in a recording would be very helpful for the research community working in the field of ecological acoustics, as it would not only reduce the time needed to be spent listening to hours of recording tapes but would also very accurately mark the presence of bird sounds in a recording. Early attempts to promote development in this area included the LifeCLEF (BirdCLEF) contest and the Bird Audio Detection (BAD) challenge [4]. In the 2017 BAD Challenge, a convolutional neural network (CNN) system reached the area under the ROC curve (AUC) accuracy of 88.7%, which was the highest score recorded, so this paper chose CNN as the basic network framework for further investigation.

Deep learning networks, as a form of neural network, have demonstrated that they can be used to extract and classify audio features, achieving more accurate results than a system using only real-recorded data [5]. However, traditional deep learning network structures require a large amount of memory and high computational power. An embedded system based on bird sound detection could actually monitor the birds in their natural environment in real time through an appropriately designed algorithm. It is thus of practical significance to port deep learning algorithms to embedded systems. However, the storage computing resources on an embedded platform are usually very limited, which makes it difficult to transplant deep learning networks into embedded mobile devices [6]. Indeed, the extensive calculations and resources involved in the operation of deep learning networks hinder the utilization of such BSD systems in mobile terminals.

The Binary Neural Networks (BNNs) can reduce the computational power and memory resources by binarizing the weight matrix and activation function of the neural network [7]. The main method to achieve this binarization is to constrain both the weight and the activation to either $+1$ or -1, which are very advantageous from a hardware perspective. Furthermore, recent studies have shown that it is possible to train a CNN on an ImageNet classification task with BinaryNet and achieve nearly state-of-the-art results [8]. During run-time, the binary convolutional neural network (BCNN) can drastically reduce the memory usage and replace most multiplications by 1-bit exclusive-not-or (XNOR) operations, which is an important advantage for hardware implementation. Nevertheless, binary networks have not yet been applied in the field of audio detection. This present study investigated the binary network as a network framework for bird sound detection to promote its implementation in BSD system hardware.

The datasets selected in this study were the development datasets from DCASE2018 Task 3. All the datasets contained 10-second-long WAV files (44.1 kHz mono PCM) and were manually labeled with a 0 or 1 to indicate the absence/presence of any birds within that 10 s audio clip. The total duration of the available recordings was approximately 91 h and the accuracy of labeling was above 95%., which makes the datasets a reliable and sufficient source for the tests. In the course of the experiments, we first extracted the desired features from the audio files, and then changed the weight of the network to $+1$ or -1 to perform the binarization operation.

Finally, the activation value of each layer of the network was binarized to complete the construction of the BCNN.

From the experimental results, it was found that the BCNN achieved a comparable performance with an ordinary CNN for bird sound detection. Further, BCNN reduced the memory requirement and reduced the hardware loss unit which is the hardware resource consumed in the calculation compared with the floating-point CNN method. Furthermore, the BCNN could also convert all 32-bit floating-point operations into 1-bit logical operations, which can improve the speed of the forward propagation. So BCNNs are of great significance to the hardware requirement in a BSD system.

The rest of the paper is organized as follows. Section 2 presents a literature review of BSD and BNNs. The method is reported in Sect. 3. Section 4 presents the results and the discussion, with the conclusions detailed in Sect. 5.

2 Literature Review

Deep learning methods, such as CNNs, have been able to achieve a 90% area under curve (AUC) accuracy on bird sound detection tasks [8]. However, deep learning networks with complex computational complexity are difficult to implement in hardware. Binary neural networks are advantageous in terms of hardware implementation. The key research on BNNs is derived from Bengio [9], who changed the weight precision into binarization while training CNNs. For BNNs, it was found that reducing the precision of the convolution weights during training (changing the floating point numbers into fixed-point numbers and dynamic fixed point numbers) only has a minimal impact on the correct rate of the neural network, but it can significantly reduce the program's memory footprint. In the subsequent study [10], Bengio's group were able to further reduce the precision of the weight into a binary value, while still achieving results similar to the highest correct rate. In addition, some methods for simplifying neural network calculations have been proposed, such as the Boolean network [11], which uses high-precision weights in training but binary weights in tests; and Plagianakos [12], which uses 3-bit integer weights. However, these BNN studies were all conducted in the field of image recognition, and there is no research in the literature related to this topic in the field of audio event detection.

Compared with the traditional deep learning networks, the memory consumption of the binary neural network model can be reduced by a factor of 32 times. BNNs can also reduce the hardware loss unit by a factor of 200 times and converts all the 32-bit floating-point operations into 1-bit logical operations. So BNNs have great advantages in model compression. At present, the BNN method has undergone some development in the field of image recognition, but has never been mentioned in the audio field, such as bird sound audio. BSD is also a project that requires suitable hardware to ensure the practicality of the deployed system. Therefore, in this paper, the BNN method was used to binarize the convolutional layer and the fully connected layer of the original CNN network, and convert the 32-bit floating-point data into $+1$ and -1 values for the BSD system. This method can not only reduce the memory needed for the network model but also improve the computational efficiency of the network in the forward propagation, which would play an important role in helping realize the application of a bird audio detection system on the hardware device.

3 Methods

The sound of birds can be easily distinguished in a piece of audio by the human ear. Therefore, this experiment used the log Mel feature as the characteristic of the audio signal, which is closer to the human auditory system. In recent years, log Mel filters have obtained good results in extracting logarithmic multi-band energy from short frames [13]. Logarithmic features contain both time domain information and frequency domain information, so they are widely used in the field of audio event detection [14–16]. In order to extract and unify the audio information before being sent to the network, we first performed feature extraction and pre-processing on the audio in the datasets. The duration of each audio file in the bird audio datasets was 10 s, but the output time was not uniform due to the differences in the recording environment and equipment. In this experiment, the feature lengths after extraction were zero-padded and cropped. When performing feature extraction, the number of sampling points was set to 2048, the number of overlaps to 720, the number of frames to 240, and the number of filters to 128. Each 10 s audio file is transformed into a 240×128-dimensional feature map and sent to the network after feature extraction. This is a relatively good performance parameter, as demonstrated in the experiments in the present study.

Although the time of the bird sounds was short, the energy distribution was more obvious. In addition, depending on the environmental conditions (such as ambient noise levels, vegetation density) and birds, the bird sounds may exhibit some local frequency shifts [17]. Therefore, the chosen neural network should be able to capture the melodic cues in the time domain and should also be robust to local frequency shifts. The convolutional filter of the CNN can analyze the multiple sub-bands of the local time of the spectrogram, which can also help preserve the local structure of the spectrogram. Therefore, the CNN was selected to learn the audio features as the basic network of BCNN. When designing the BCNN network, herein the structure was adjusted to place the BatchNormal layer, which can normalize the data, and the binarization activation layer before the convolution operation. The purpose of this change was to prevent the result of the activation layer from becoming only one value after pooling. The network structures are compared in Fig. 1.

The binarization in the CNN is mainly performed in two steps:

(1) In the forward propagation process, the weight w of the CNN is binarized according to the sign function. The specific calculation formula is as follows:

$$w^b = Sign(w) = \begin{cases} +1, & w \geq 0 \\ -1, & w < 0 \end{cases} \tag{1}$$

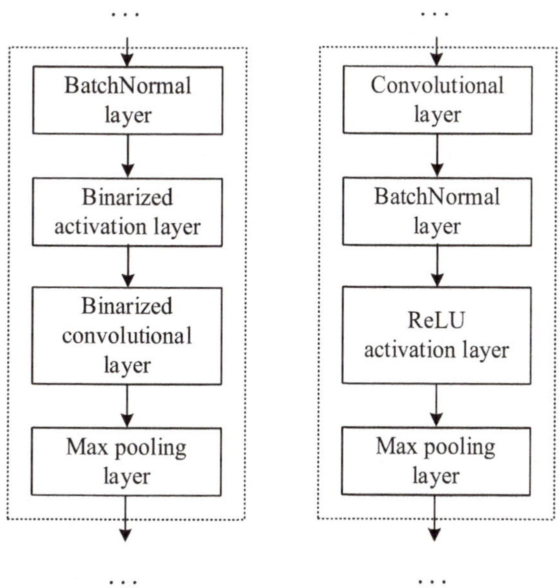

Fig. 1 The binarized convolution block (left) and the non-binarized convolution block (right) structures

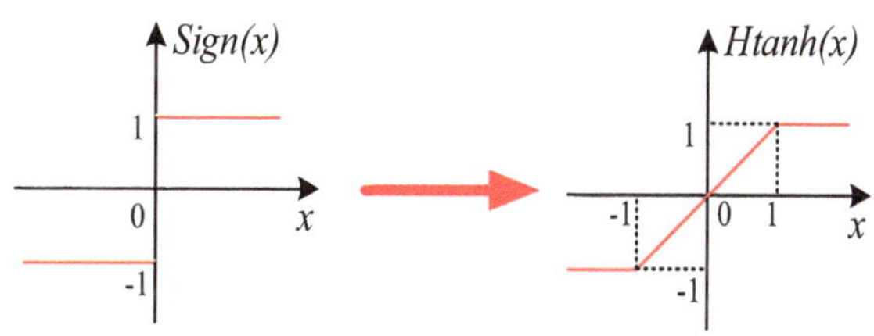

Fig. 2 Sign(x) and Htanh(x) functions

where w^b is the value of the weight w after binarization; whereupon, since only two values are allowed in the binarization, w^b is taken as 1 when $w = 0$.

(2) After the weight is further binarized, the activation value then needs to be further binarized. This paper implements the binarization of the activation value by adding *Sign(x)* as the activation function. The formula for the activation value is as follows:

$$X_i^b = L\left(X_{(i-1)}\right) = sign\left(X_{(I-1)}\right) \tag{2}$$

where X_i^b is the input value of the ith layer of the binarized network, and $L\left(X_{(i-1)}\right)$ is the value obtained by the ith layer binarization activation layer.

The gradient of the *Sign(x)* function is zero, which tends to cause the gradient to disappear during back propagation. To solve this problem, BCNN uses the *Htanh(x)* function (Fig. 2) instead of the *Sign(x)* function [18] as the activation function. The images of the two functions and their conversion relationships are shown in Fig. 2.

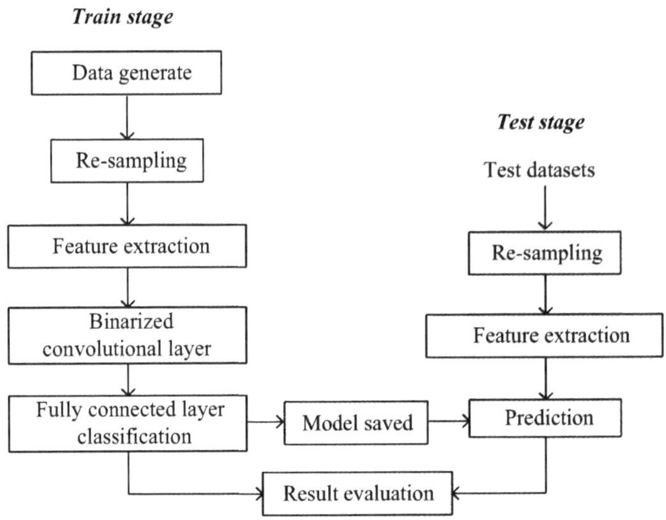

Fig. 3 Flowchart of the designed system

4 Results

In this paper, the task of building an audio event detection system was mainly carried out for task 3 of DCASE2018. The task was to design a system that, given a short audio recording, returns a binary decision for the presence/absence of bird sound (bird sound of any kind). The output can be just binary, but weighted/probability outputs were encouraged in the continuous range [0,1] for the purposes of evaluation. For the main assessment, this paper uses the well-known "Area Under the Receiver Operating Characteristic (ROC) Curve" (AUC) measure of classification performance. ROC is a comprehensive indicator reflecting the continuous variables of sensitivity and specificity. Each point on the roc curve reflects the sensitivity to the same signal stimulus.

The overall design process of this system is shown in Fig. 3.

4.1 Experimental Processing of the Datasets

The datasets selected in the present study were the development datasets of DCASE2018 Task 3, which were from three different environments and comprised 35,800 audio files. In order to improve the adaptability of the network structure, the datasets from the three different sources were mixed and processed, and all 35,800 audio files were randomly divided into the train set, test set, and validation set. The

Table 1 The network hyperparameters

	Hyperparameters
Convolutional layers	4
Filter shape	5-by-5
Pool size	(2,2)
Fully connected layers	2

distribution ratio of these files was 8:1:1. Finally, a three-fold cross-validation was performed to verify whether the model was overfitting.

Further, the three datasets were divided into "folds" for cross-validation, with a stratified 3-way cross-validation performed. In each fold, two sets were used for training and the other one for testing. This allowed the network to study the way the system behaves when exposed to data from unseen conditions, just like in the final challenge evaluation.

4.2 Results of the BCNN and CNN in Bird Detection

In the present study, a binarized convolutional neural network system applied for a bird sound detection system was built under the framework of Pytorch [19]. Pytorch can design the activation value of each layer as a custom function, which gives powerful flexibility to the building of a binarized neural network. Here, the BCNN network structure was constructed by changing the weight and activation value of the convolution layer in the Pytorch framework.

The network structure used in this study consisted of 4 convolutional layers and 3 fully connected layers. The number of feature maps per convolutional layer was {128, 64, 128, 128}, the three fully connected layers was {128, 50, 1}, and the last layer was the sigmoid output layer. The input signal was a Log Mel feature, which is closer to the human auditory system. The network configuration hyperparameters are given in Table 1.

After the system was set up, the binarized weight and activation value could be obtained by printing the weight and the activation value. The weight value could become $+1$ and -1 after passing through the binarized convolution layer. The Table 2 shows the performance of the BCNN for the bird sound detection, and also the results of the common CNN under the same parameters.

From the above results, it can be seen that BCNN achieves 82.90% of the performance of the same parameter CNN on the test AUC score. Furthermore, the parameter Test AUC score/Train AUC score was chosen to compare the difference in the performance of the network between the training datasets and the test datasets. It is generally believed that the greater value of Test AUC score/Train AUC score, the better the adaptive ability the network to the new datasets. From the result of $\frac{Test\ AUC\ Score}{Train\ AUC\ Score}$, it can be seen that BCNN achieved a stronger generalization per-

Table 2 The AUC scores of CNN and BCNN on BAD

	Train AUC score (%)	Verify AUC score (%)	Test AUC score (%)	$\frac{Test\ AUC\ Score}{Train\ AUC\ Score}$ (%)
CNN	99.89	92.31	82.75	82.84
BCNN	72.14	70.51	68.60	95.09
$\frac{BCNN}{CNN}$	72.22	76.38	82.90	

formance on the new data set than the ordinary CNN (95.09% vs. 82.84%). More importantly, the use of BCNN reduced the memory requirement by a factor of 32 times and reduced the hardware loss unit by a factor of 200 times compared with the floating-point CNN method. BCNN also converts all 32-bit floating-point operations into 1-bit logical operations, which not only improves the hardware prediction speed but also promotes the hardware implementation of the bird sound detection system.

5 Conclusion

In order to solve the challenge of implementing bird sound detection based on deep learning networks in hardware, this paper proposes a method based on the binarized convolutional neural network for a BSD system. This method achieved an AUC score of 68.60% in the BAD task of DCASE 2018, which was comparable with the performance of CNN on the unseen evaluation data. BCNN not only displayed better generalization performance than CNN but also is less computational and resource expensive as it simplifies the multiplication and addition operations in the convolution into a simple binary logic operation. These characteristics of BCNN are important for the porting of BSD systems on mobile hardware.

Acknowledgements This work is partially supported by Youth Innovation Projects of Beijing University of Posts and Telecommunications (2017RC16): The method of evaluating the performance of FPGA computation platform for deep learning systems.

References

1. Emre C, Sharath A, Giamabattista P, Konstantinos D, Tuomas V (2017) Convolutional recurrent neural networks for bird audio detection. 25th Europe Signal Process (EUSIPCO) 2017. pp 1729–1733
2. Tiago AM, Len T, Martin S, David M (2013) Estimating animal population density using passive acoustics. Biol Rev 88(2):287–309
3. Abraham LB, Matthew WM, Joshua TA, Collin AE, Bernie RT, Donald AC (2014) Vocal activity as a low cost and scalable index of seabird colony size. Conserv Biol 28(4):1100–1108
4. Dan S, Mike W, Yannis S, Herve G (2016) Bird detection in audio: a survey and a challenge. In: International workshop on machine learning for signal processing (MLSP). pp 13–16

5. Yasutaka N, Masahiro S, Taisuke N, Norihito S, Toshiya O, Nobutaka O (2016) DNN-based environmental sound recognition with real-recorded and artificially-mixed training data. Inter-Noise 3164–3173
6. Chao L, Zhiyong Z, Dong W (2014) Pruning neural networks by optimal brain damage. Int Speech Commun Assoc (INTERSPEECH) 1092–1095
7. Om PP, Aruna T (2015) Advance quantum based binary neural network learning algorithm. Network Parallel/Distrib Comput (SNPD) 1–6
8. Dan S, Yannis S, Mike W, Hanna P (2018) Automatic acoustic detection of birds through deep learning. The first bird audio detection challenge. Meth Ecol Evol 15:1–21
9. Mohammad R, Vicente O, Joseph R, Ali F(2016) XNOR-Net: ImageNet classification using binary convolutional neural networks. Europ Conf Comput Vision (ECCV) 1–17
10. Matthieu C, Yoshua B, Jean-Pierre D (2015) BinaryConnect: training deep neural networks with binary weights during propagations. Neural Inform Process Syst (NIPS) 2:3123–3131
11. Minje K, Paris S (2015) Bitwise neural networks. Int Conf Mach Learn (ICML) 37
12. Plagianakos VP, Vrahatis MN (1999) Training neural networks with 3-bit integer weights. Genet Evol Comput Conf (GECCO) 1:910–915
13. Emre C, Toni H, Heikki H, Tuomas V (2015) Polyphonic sound event detection using multi label deep neural networks. Int Joint Conf Neural Network (IJCNN) 1–7
14. Adam C, Brody H, Tao W, David JW, Andrew YN, Bryan C (2013) Deep learning with COTS HPC systems. Int Conf Machine Learn (ICML) 28:1337–1345
15. Song H, Jeff P, John T, William JD (2015) Learning both weights and connections for efficient neural networks. Neural Inform Process Syst (NIPS) 1:1135–1143
16. Steve KE, Rathinakumar A, Paul AM, John VA, Dharmendra SM (2015) Backpropagation for energy-efficient neuromorphic computing. Neur Inform Process Syst (NIPS) 1:1117–1125
17. Elizabeth PD (2009) Ecology shapes birdsong evolution: variation in morphology and habitat explains variation in WhiteCrowned Sparrow song. Am Natural 74(1):24–33
18. Matthieu C, Jean-Pierre D, Yoshua B (2015) Training deep neural networks with low precision multiplications. Int Conf Learn Represent (ICLR)
19. Huth J, Timothee M, Angelo A (2018) Convis: a toolbox to fit and simulate filter-based models of early visual processing. Front Neuroinfor. https://doi.org/10.3389/fninf.2018.00009. Last Assessed 7 Mar 2018

Adaptive Consistent Dictionary Learning for Audio Declipping

Penglong Wu, Xia Zou, Meng Sun, Li Li and Xingyu Zhang

Abstract Clipping is a common problem in audio processing. Clipping distortion can be solved by the recently proposed consistent Dictionary Learning (cDL), but the performance of restoration will decrease when the clipping degree is large. To improve the performance of cDL, a method based on adaptive threshold is proposed. In this method, the clipping degree is estimated automatically, and the factor of the clipping degree is adjusted according to the degree of clipping. Experiments show the superior performance of the proposed algorithm with respect to cDL on audio signal restoration.

Keywords Audio declipping · Dictionary learning · Adaptive threshold · Adaptive consistent dictionary learning

1 Introduction

Due to the influence of recording equipment and recording environments, signals may be distorted during the first stage of signal recording. Among the distortions, clipping is a common type. Clipping will impart harsh noise to the signal. If the degree of clipping is large, the quality of signals will be seriously affected, which

P. Wu · X. Zou (✉) · M. Sun · L. Li · X. Zhang
Army Engineering, University of PLA, 210000 Nanjing, China
e-mail: zlc1997@163.com

P. Wu
e-mail: 17551050128@163.com

M. Sun
e-mail: sunmengccjs@163.com

L. Li
e-mail: 754583546@qq.com

X. Zhang
e-mail: zxynbnb@126.com

W. Li et al. (eds.), *Proceedings of the 6th Conference on Sound and Music Technology (CSMT)*, Lecture Notes in Electrical Engineering 568,
https://doi.org/10.1007/978-981-13-8707-4_7

is detrimental to their further processing. Declipping is an operation that aims to eliminate the clipping distortion.

Early strategies to recover a clipped signal include the use of autoregressive modeling [1], bandwidth limited models [2], and Bayesian estimation [3]. Recently, sparsity-based declipping techniques have attracted widespread attention. The basic idea of the sparsity method is that signals can be sparsely represented using a dictionary. Based on this idea, declipping can be treated as solving a sparse decomposition problem on the unclipped samples [4]. The reconstruction of signals can be greatly improved by considering additional useful information. Given the fact that the amplitude of clipped samples should be larger than the amplitude threshold, sparse decomposition with amplitude constraints was proposed in [5–7] and solved using the two-step algorithm or the convex optimization toolbox. These approaches used the constrained Orthogonal Matching Pursuit (cOMP) algorithm, which is an improved version of OMP that imposes constraints on the optimization variables. In the first stage, cOMP discards unreliable samples from the data and attempts to detect the optimal basis vectors using only reliable samples. In the second stage, clipping constraints are imposed to improve the reconstruction. Although the cOMP algorithm helps restore the clipped signal, one drawback of it is that, because it discards unreliable samples, the first stage does not consider the information stored in the clipped samples, which will affect the accuracy of reconstruction. Another drawback is that cOMP's computational complexity is high.

To overcome these drawbacks, smooth regularizers that lead to simple, unconstrained cost functions were proposed in [8]. These can be optimized using variants of Iterative Hard Thresholding (IHT). In order to improve the signal reconstruction, other useful information such as the perception information from humans [6], multichannel data [9], and social sparsity priors [10] has also been used.

The methods described above are all based on sparse decomposition by fixed dictionaries. Thus, choosing a suitable dictionary is crucial to the performance of the reconstruction. Therefore, the consistent Dictionary Learning (cDL) algorithm, which is able to learn useful information directly from nonlinearly clipped measurements, was proposed in [11]. Since dictionary learning usually alternates between updating bases and their weights, it requires calculations that are straightforward and robust to initializations. So, a soft data-consistency metric that provides a simple optimization problem for dictionary learning was proposed in [11]. The proposed algorithm alternates between a sparse coding step and a dictionary update step. At the sparse coding step, a consistent Iterative Hard Thresholding (cIHT) algorithm was used to sparsely decompose the signal on the overcomplete Discrete Cosine Transform (DCT) dictionary to obtain sparse coefficients. At the dictionary update step, sparse coefficients were used to update the dictionary. After several iterations, a new dictionary and sparse coding coefficients can be obtained.

Although the cDL algorithm produces positive reconstruction performance when signals are mildly clipped, for signals that are severely clipped, ideal reconstruction performance cannot be achieved using the cDL algorithm. To solve this problem, we propose in this paper the adaptive consistent Dictionary Learning (acDL) algorithm, which is not only able to learn directly from nonlinearly clipped measurements but is also able to use information on the clipping degree of signals to improve reconstruction performance. The main difference between the cDL and acDL algorithms is that acDL algorithm can use information on the clipping degree to overcome the shortcoming of the cDL algorithm, which does not produce good reconstruction performance when signals are severely clipped. Experiments show that reconstruction performance can be improved significantly by using clipping degree information.

2 Clipping Model

When a digital signal is clipped, all its amplitude values lying beyond a maximum level or below a minimum level are limited inside a specific interval, as shown in Fig. 1.

Let $\boldsymbol{x} \in \mathbb{R}^N$ be a clean signal, and $\boldsymbol{y} \in \mathbb{R}^N$ be a clipped signal. Each sample y_i can be expressed as

$$y_i = \begin{cases} \theta^+ & x_i \geq \theta^+ \\ \theta^- & x_i \leq \theta^- \\ x_i & \theta^- < x_i < \theta^+ \end{cases} \tag{1}$$

where θ^+, θ^- are the positive and negative clipping thresholds, and x_i is the sample of the clean signal. N is the total number of samples. From (1), we can define different sets of samples. Let Ω_r be the set of reliable samples that have no clipping. Let $\Omega_{\mathrm{c}+}$ be the set of positive clipping samples and $\Omega_{\mathrm{c}-}$ be the set of negative clipping samples. These can be written as: $\Omega_\mathrm{r} = \{i \in [1, 2, \ldots, N] : \theta^- < x_i < \theta^+\}$, $\Omega_{\mathrm{c}+} = \{i \in [1, 2, \ldots, N] : x_i > \theta^+\}$, and $\Omega_{\mathrm{c}-} = \{i \in [1, 2, \ldots, N] : x_i < \theta^-\}$. With these sets of samples, (1) can be expressed as

$$\boldsymbol{y} = \boldsymbol{M}_{\Omega_\mathrm{r}}\boldsymbol{x} + \theta^+\boldsymbol{M}_{\Omega_{\mathrm{c}+}}1 + \theta^-\boldsymbol{M}_{\Omega_{\mathrm{c}-}}1 \tag{2}$$

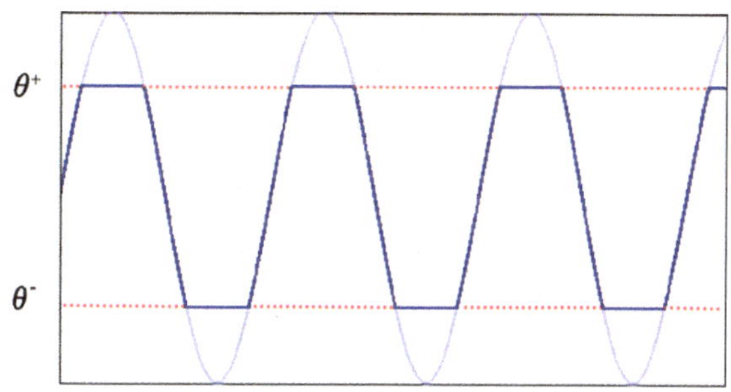

Fig. 1 Signal clipping

where $1 \in \mathbb{R}^N$ is the all-ones vector with some specific length, and $\boldsymbol{M}_{\Omega_r}$, $\boldsymbol{M}_{\Omega_{c+}}$ and $\boldsymbol{M}_{\Omega_{c-}}$ are the diagonal masking matrix in $\{0, 1\}^{N \times N}$ where $[\boldsymbol{M}_S]_{i,i} = 1$ if $i \in S$, or 0 otherwise.

3 Dictionary Learning

3.1 Consistent Dictionary Learning

As mentioned in paper [11], declipping was reformulated as a problem of minimizing the distance between the approximated signal and the feasible set as follows:

$$\min_{D \in D, \boldsymbol{a}_t} \sum_t \mathrm{d}\left(\boldsymbol{D}\boldsymbol{a}_t, \ell(\boldsymbol{y}_t)\right)^2 \quad s.t. \forall t, \|\boldsymbol{a}_t\|_0 \le K \tag{3}$$

where $\{x_t\}_{1 \ldots T}$ is a collection of T signals. To avoid scaling ambiguity, the dictionary is often constrained to be in $D = \{\boldsymbol{D} \in \mathbb{R}^{N \times M} | \forall i, \|\boldsymbol{d}_i\|_2 < 2\}$ as proposed in [12]. K is the sparsity, and $\ell(\boldsymbol{y})$ is a feasible set defined as

$$\ell(\boldsymbol{y}) \triangleq \left\{\boldsymbol{x} \middle| \boldsymbol{M}_{\Omega_r}\boldsymbol{y} = \boldsymbol{M}_{\Omega_r}\boldsymbol{x}, \boldsymbol{M}_{\Omega_{c+}}\boldsymbol{x} \ge \boldsymbol{M}_{\Omega_{c+}}\boldsymbol{y}, \boldsymbol{M}_{\Omega_{c-}}\boldsymbol{x} \le \boldsymbol{M}_{\Omega_{c-}}\boldsymbol{y}\right\} \tag{4}$$

and $\mathrm{d}\left(\boldsymbol{D}\boldsymbol{a}_t, \ell(\boldsymbol{y}_t)\right)$ is the Euclidean distance defined as

$$\mathrm{d}\left(\boldsymbol{D}\boldsymbol{a}_t, \ell(\boldsymbol{y}_t)\right) = min_{z \in \ell(\boldsymbol{y}_t)} \|\boldsymbol{D}\boldsymbol{a}_t - z\|_2 \tag{5}$$

Therefore, using (3), (4), and (5), the problem of declipping can be reformulated as

$$\begin{aligned} &\min_{D \in D, \boldsymbol{a}_t} \sum_t \left\|\boldsymbol{M}_{\Omega_r}(\boldsymbol{y} - \boldsymbol{D}\boldsymbol{a}_t)\right\|^2 + \left\|\boldsymbol{M}_{\Omega_{c+}}\left(\theta^+ 1 - \boldsymbol{D}\boldsymbol{a}_t\right)_+\right\|^2 \\ &+ \left\|\boldsymbol{M}_{\Omega_{c-}}\left(\theta^- 1 - \boldsymbol{D}\boldsymbol{a}_t\right)_-\right\|^2 \quad \text{s.t.} \|\boldsymbol{a}_t\|_0 \le K \end{aligned} \tag{6}$$

where $(u)_+ = \max(0, u)$ and $(u)_- = \min(0, u)$. The first term of (6) promotes the candidate signal to match the observed clipped signal for the unclipped samples. Thanks to the thresholding functions (i.e., $(u)_+$ and $(u)_-$), the second term enforces the reconstructed signal to be higher than θ^+ on the set of Ω_{c+}, and the third term enforces the reconstructed signal to be lower than θ^- on the set of Ω_{c-}. All three terms fit well with the goal of declipping.

3.2 Adaptive Consistent Dictionary Learning

The cDL algorithm has positive impacts on the restoration of clipped signals, but as the clipping degree increases, the declipping performance will decline. The reason for this is, while the second (or third) term of (6) makes the amplitude of the reconstructed signal higher (or lower) than the clipping thresholds in the clipped domain, it does not consider how much higher or lower it should be.

The true amplitude of a clipped sample is usually different from the clipping threshold. Using only the information that the amplitude of the reconstructed signal is bigger (positive clipping) or smaller (negative clipping) than the clipping threshold does not model this fact very well. Considering that the greater the extent of clipping, the bigger (positive clipping) or smaller (negative clipping) the value between the amplitude of the clean signal and the clipping threshold. Therefore, inspired by this intuitive rule and cDL, we propose an adaptive consistent Dictionary Learning algorithm in this paper. This algorithm strengthens the consistency of the reconstructed signal with the clipped observations by adding a factor of clipping degree. So, the problem of declipping can be re-written as

$$\min_{D \in \mathcal{D}, \boldsymbol{a}_t} \sum_t \left\| \boldsymbol{M}_{\Omega_r}(\boldsymbol{y} - \boldsymbol{D}\boldsymbol{a}_t) \right\|^2 + \left\| \boldsymbol{M}_{\Omega_{c+}} [\gamma(k)(\theta^{+}1 - \boldsymbol{D}\boldsymbol{a}_t)]_{+} \right\|^2 + \left\| \boldsymbol{M}_{\Omega_{c-}} \left[\gamma(k)\left(\theta^{-}1 - \boldsymbol{a}_t\right)\right]_{-} \right\|^2 \quad \text{s.t.} \|\boldsymbol{a}_t\|_0 \le K \tag{7}$$

where $\gamma(k)$ is the extent-of-clipping factor, $k = N_c/N$ the ratio of the number of clipped samples to the total number of samples, N_c the number of clipped samples, and N the total number of samples.

Equation (7) is obtained by adding $\gamma(k)$ to the second and third terms of (6). By introducing $\gamma(k)$ into the optimization problem, (7) uses not only the clipping threshold but also the clipping degree information. In general, the more severely clipped the signal is, the more clipped samples there are. Here k is used to represent the clipping degree information. In the experimental stage, we evaluate 10 speech signals and 10 music signals. Every speech signal is artificially clipped by various degrees. For different degrees of clipping, we estimate the best γ that can fit the data pairs of γ and k well. To intuitively and clearly show the relationship between γ and k, $\ln(\gamma)$ and $\ln(k)$ are taken as axes, as illustrated in Fig. 2.

For audio signals with different clipping degrees, we expect that the algorithm can adjust γ adaptively and that the adjusted γ can be used to (7). Therefore, an exponential function is adopted to fit $\ln(k)$ and $\ln(\gamma)$, as depicted in Fig. 3. The fitting function is $\ln(\gamma) = 2.13 \times e^{0.4254 \times \ln(k)}$.

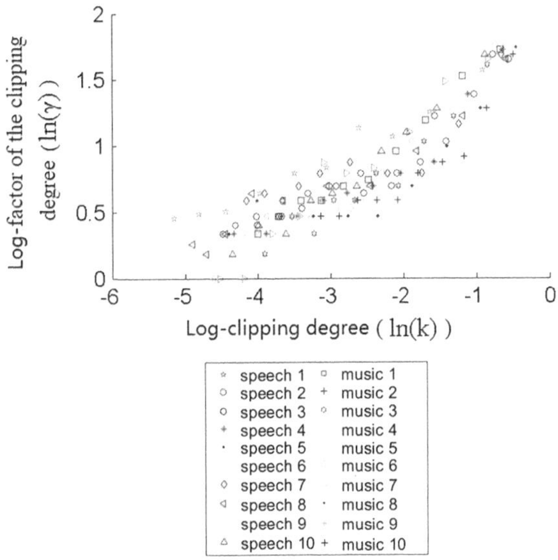

Fig. 2 Scatter diagram. The different mark shapes represent different audios

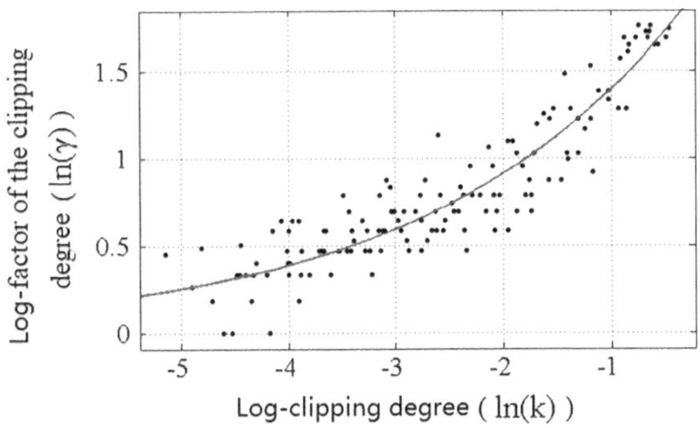

Fig. 3 Diagram of scatter data and fitting function

3.3 acDL Algorithm

The proposed algorithm shown in Algorithm 1, which is obtained by referring to [11], alternates between a sparse coding step and a dictionary update step. Gradient descent is used at both steps.

Algorithm 1. Adaptive consistent Dictionary Learning algorithm

Input: $\{\boldsymbol{y}_t\}_{1\ldots T}, \boldsymbol{D}_0, \boldsymbol{a}_0, n_1, n_2, \mu_1, \mu_2$

initialize: $\boldsymbol{D} = \boldsymbol{D}_0, \boldsymbol{a} = \boldsymbol{a}^0$
for $t = 1$ to T do
for $i = 1$ *to* n_1 do
$\boldsymbol{a}_t = H_K(\boldsymbol{a}_t + \mu_1 \boldsymbol{D}^T \beta(\boldsymbol{y}_t - \boldsymbol{D}\boldsymbol{a}_t))$
for $j = 1$ *to* n_2 do
$\boldsymbol{D} = \Pi_{\boldsymbol{D}}(\boldsymbol{D} + \mu_2 \sum_t \beta(\boldsymbol{y} - \boldsymbol{D}\boldsymbol{a}_t)\boldsymbol{a}_t^T)$

Return: $\widehat{\boldsymbol{D}}, \{\widehat{\boldsymbol{a}}_t\}_{1\ldots T}$

where $\boldsymbol{D}_0$ is the initial dictionary, which is a DCT dictionary in this algorithm, and $\boldsymbol{a}^0$ is the initial sparse coding vector, which is an all-zeros vector. n_1, n_2 are the numbers of the gradient descent step, and μ_1, μ_2 are the step sizes, respectively. $H_K(x)$ is the K-term thresholding operator setting to zero all but the K highest-magnitude coefficients of x, and $\Pi_{\boldsymbol{D}}(x)$ is the Euclidean projection of x onto $\boldsymbol{D}$. And $\beta(x) = \boldsymbol{M}_{\Omega_r} x + \boldsymbol{M}_{\Omega_{c+}} \cdot [\gamma(k) \cdot x]_+ + \boldsymbol{M}_{\Omega_{c-}} \cdot [\gamma(k) \cdot x]_-$.

4 Experiment

Two speech signals and two music signals were used to evaluate the performance of the algorithm proposed in this paper. The signals were both sampled at 16 kHz. In the preprocessing stage, Hamming windows were selected as the processing window and the DCT dictionary, which has 512 atoms, was selected for the initialization dictionary. The signal distortion ratio (SDR) and PESQ methods were used to evaluate the quality of the reconstructed signals.

4.1 *Signal Distortion Ratio (SDR)*

The SDR of the reconstructed signal is defined as

$$\mathrm{SDR} = 10 \log \frac{\left\| \left(\boldsymbol{M}_{\Omega_{c+}} + \boldsymbol{M}_{\Omega_{c-}} \right) \cdot \boldsymbol{x} \right\|^2}{\left\| \left(\boldsymbol{M}_{\Omega_{c+}} + \boldsymbol{M}_{\Omega_{c-}} \right) \cdot (\boldsymbol{x} - \hat{\boldsymbol{x}}) \right\|^2} \tag{8}$$

where $\hat{x}$ is the reconstructed signal.

$$\Delta\mathrm{SDR} = 10 \log \frac{\left\| \left(\boldsymbol{M}_{\Omega_{c+}} + \boldsymbol{M}_{\Omega_{c-}} \right) \cdot \boldsymbol{x} \right\|^2}{\left\| \left(\boldsymbol{M}_{\Omega_{c+}} + \boldsymbol{M}_{\Omega_{c-}} \right) \cdot (\boldsymbol{x} - \hat{\boldsymbol{x}}) \right\|^2}$$

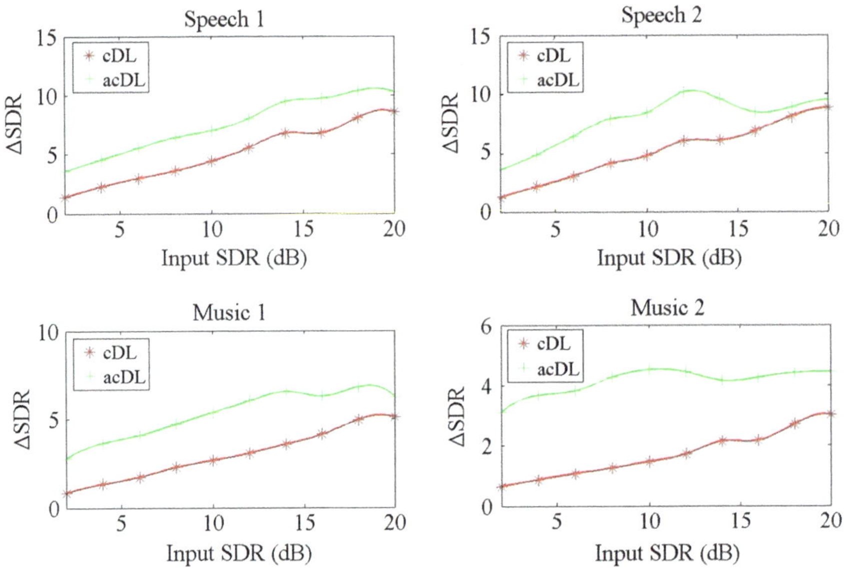

Fig. 4 Comparison of performance of acDL and cDL algorithms

$$-10\log\frac{\left\|\left(\boldsymbol{M}_{\Omega_{c+}}+\boldsymbol{M}_{\Omega_{c-}}\right)\cdot\boldsymbol{x}\right\|^2}{\left\|\left(\boldsymbol{M}_{\Omega_{c+}}+\boldsymbol{M}_{\Omega_{c-}}\right)\cdot(\boldsymbol{x}-\boldsymbol{y})\right\|^2} \quad (9)$$

ΔSDR, defined in (9), is the difference between input SDR and output SDR. It reflects the relative improvements from the input to the output.

Figure 4 shows the performance of the adaptive consistent Dictionary Learning (acDL) and consistent Dictionary Learning (cDL) algorithms. To compare the two algorithms comprehensively, the reconstruction performance at different clipping levels, ranging from severe clip (SDR = 2dB) to mild clip (SDR = 20dB), is given in Fig. 4. It can be seen from Fig. 4 that, whether the signal is severely or mildly clipped, the performance of the proposed algorithm is significantly better than that of cDL. Also, the acDL's use of clipping degree information to enhance the consistency of the reconstructed signal with the clipped observations improves reconstruction, as observed in Fig. 5.

Figure 5 also shows that, for both speech and music, the proposed algorithm can reconstruct waveforms that are similar to their counterparts. The figure also shows that the amplitude of the sample points is closer to the amplitude of the original samples. This also illustrates that, compared to cDL, the SDRs of the signals reconstructed by acDL have been greatly improved.

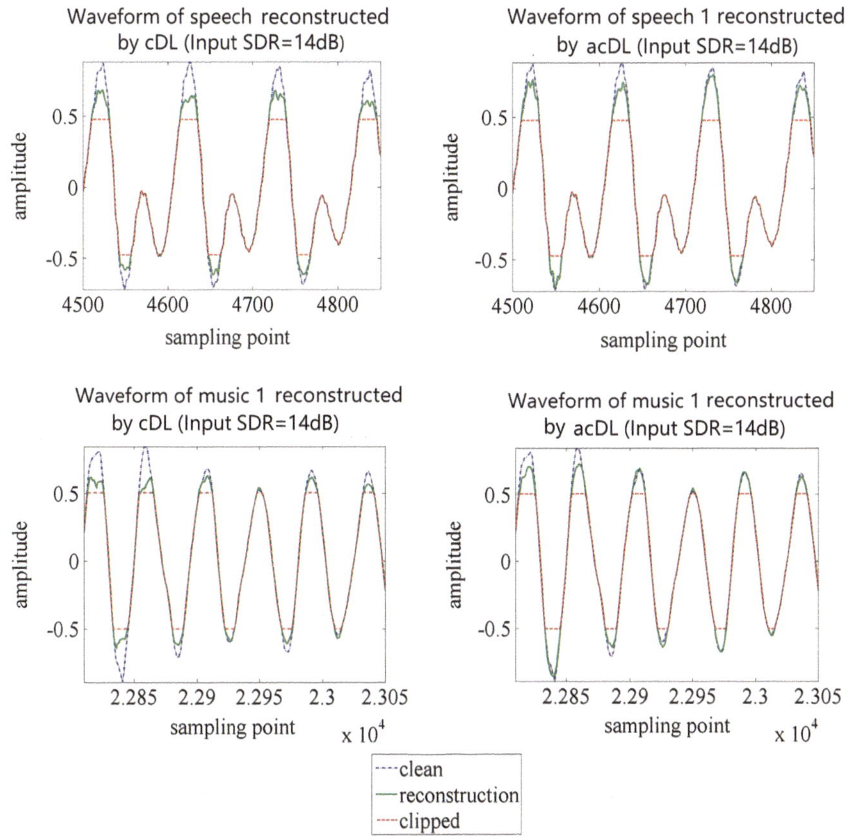

Fig. 5 Waveform of reconstructed signals

4.2 PESQ

In this paper, the two speech signals were artificially clipped to varying degrees, and signals for each degree of clipping were reconstructed by cDL and acDL. Finally, the reconstructed speech signals were scored by PESQ. The resulting PESQ values are shown in Tables 1 and 2.

Tables 1 and 2 show the PESQ scores of the speech signals with different degrees of clipping as reconstructed by cDL and acDL. The table data illustrates that, for the different speech signals reconstructed by acDL, the PESQ scores are significantly improved over cDL. Compared to cDL, for both slightly and severely clipped signals, the improvement in PESQ scores achieved by using acDL is significant.

Table 1 PESQ scores of speech 1

Input SDR(dB)	PESQ of clipping speech	PESQ of speech reconstructed by cDL	PESQ of speech reconstructed by acDL
2	1.73	2.05	2.19
4	2.10	2.54	2.64
6	2.35	2.78	2.89
8	2.57	3.00	3.07
10	2.72	3.23	3.32
12	2.90	3.41	3.53
14	3.08	3.54	3.74
16	3.19	3.61	3.82
18	3.32	3.74	3.94
20	3.45	3.85	3.95

Table 2 PESQ scores of speech 2

Input SDR(dB)	PESQ of clipping speech	PESQ of speech reconstructed by cDL	PESQ of speech reconstructed by acDL
2	1.95	2.16	2.39
4	2.33	2.73	3.05
6	2.56	3.12	3.32
8	2.84	3.32	3.52
10	3.02	3.48	3.65
12	3.16	3.69	3.90
14	3.27	3.79	3.97
16	3.37	4.06	4.08
18	3.45	4.04	4.17
20	3.51	4.16	4.26

5 Conclusion

In this paper, we proposed an audio declipping algorithm that can adapt to different degrees of clipping. This algorithm shows excellent performance when audio declipping with different clipping degrees. It has bettered the reconstruction performance under typical clipping, and it adapts well to clipping situations in real environments. Compared to cDL, the algorithm proposed in this paper shows better performance.

Acknowledgements Thanks are due to Mr. Zou for assistance with the experiments and to Mr. Sun for valuable discussion. This paper is supported by The National Natural Science Foundation of China (61471394) and The National Natural Foundation of Jiangsu Province for Excellent Young Scholars (BK20180080).

References

1. Janssen AJEM, Veldhuis R., Vries L (1986) Adaptive interpolation of discrete-time signals that can be modelled as autoregressive processes. IEEE Trans Acoust Speech Signal Process 34(2):317–330
2. Abel JS, Smith JO (1991) Restoring a clipped signal. In: International conference on acoustics, speech, and signal processing. IEEE, pp 1745–1748
3. Godsill SJ, Wolfe PJ, Fong WN (2001) Statistical model-based approaches to audio restoration and analysis. J New Music Res 30(4):323–338
4. Adler A, Emiya V, Jafari MG, Elad M, Gribonval R, Plumbley MD (2012) Audio inpainting. IEEE Trans Audio Speech Lang Process 20(3):922–932
5. Adler A, Emiya V, Jafari MG, Elad M, Gribonval R, Plumbley MD (2011) A constrained matching pursuit approach to audio declipping. In: IEEE international conference on acoustics, speech and signal processing. IEEE, pp 329–332
6. Defraene B, Mansour N, De Hertogh S, Van Waterschoot T, Diehl M, Moonen M (2013) Declipping of audio signals using perceptual compressed sensing. IEEE Trans Audio Speech Lang Process 21(12):2627–2637
7. Foucart S, Needham T (2016) Sparse recovery from saturated measurements. Inf Infer A J IMA 6(2):196–212
8. Kitic S, Jacques L, Madhu N, Hopwood MP, Spriet A, De Vleeschouwer C (2013) Consistent iterative hard thresholding for signal declipping. In: IEEE international conference on acoustics, speech and signal processing. IEEE, pp 5939–5943
9. Ozerov A, Bilen Ç, Pérez P (2016) Multichannel audio declipping. In: IEEE international conference on acoustics, speech and signal processing. IEEE, pp 659–663
10. Siedenburg K, Kowalski M, Dörfler M (2014) Audio declipping with social sparsity. In: IEEE international conference on acoustics, speech and signal processing. IEEE, pp 1577–1581
11. Rencker L, Bach F, Wang W, Plumbley MD (2018) Consistent dictionary learning for signal declipping. In: International conference on latent variable analysis and signal separation. Springer, Cham, pp 446–455
12. Mairal J, Bach F, Ponce J (2014) Sparse modelling for image and vision processing. Found Trends® Comput Graph Vision 8(2–3):85–283

A Comparison of Attention Mechanisms of Convolutional Neural Network in Weakly Labeled Audio Tagging

Yuanbo Hou, Qiuqiang Kong and Shengchen Li

Abstract Audio tagging aims to predict the types of sound events occurring in audio clips. Recently, the convolutional recurrent neural network (CRNN) has achieved state-of-the-art performance in audio tagging. In CRNN, convolutional layers are applied on input audio features to extract high-level representations followed by recurrent layers. To better learn high-level representations of acoustic features, attention mechanisms were introduced to the convolutional layers of CRNN. Attention is a learning technique that could steer the model to information important to the task to obtain better performance. The two different attention mechanisms in the CRNN, the Squeeze-and-Excitation (SE) block and gated linear unit (GLU), are based on a gating mechanism, but their concerns are different. To compare the performance of the SE block and GLU, we propose to use a CRNN with a SE block (SE-CRNN) and a CRNN with a GLU (GLU-CRNN) in weakly labeled audio tagging and compare these results with the CRNN baseline. The experiments show that the GLU-CRNN achieves an area under curve score of 0.877 in polyphonic audio tagging, outperforming the SE-CRNN of 0.865 and the CRNN baseline of 0.838. The results show that the performance of attention based on GLU is better than the performance of attention based on the SE block in CRNN for weakly labeled polyphonic audio tagging.

Keywords Audio tagging · Convolutional recurrent neural network (CRNN) · Convolutional neural network (CNN) · Squeeze-and-Excitation (SE) block · Gated linear unit (GLU)

Y. Hou (✉) · S. Li
Beijing University of Posts and Telecommunications, Beijing, People's Republic of China
e-mail: hyb@bupt.edu.cn

S. Li
e-mail: shengchen.li@bupt.edu.cn

Q. Kong
Centre for Vision, Speech and Signal Processing, University of Surrey, Guildford, UK
e-mail: q.kong@surrey.ac.uk

W. Li et al. (eds.), *Proceedings of the 6th Conference on Sound and Music Technology (CSMT)*, Lecture Notes in Electrical Engineering 568,
https://doi.org/10.1007/978-981-13-8707-4_8

1 Introduction

Audio tagging is an audio annotation task that aims to perform multilabel audio classification on audio clips. Audio tagging has many applications such as audio information retrieval, audio classification [1], acoustic scene recognition [2] and industry sound recognition [3].

Accurate audio tagging relies on the amount of labeled audio data, including strongly and weakly labeled audio data. In strongly labeled data [1], each audio chunk is labeled with both audio tags and the occurrence time of corresponding sound events. However, since strongly labeled data is scarce and insufficient for industrial use, the audio research community has turned to large-scale audio datasets without the accurate occurrence time of sound events, which is referred to as weakly labeled data [4]. Weakly labeled audio tagging is a newly proposed task in the Detection and Classification of Acoustic Scenes and Events (DCASE) 2017 challenge [5]. However, although the amount of weakly labeled data is huge, the data are difficult to use because in weakly labeled data, the tags of sound events only indicate the presence but not the occurrence times of the events. Moreover, these audio recordings are from real life, the duration of sound events varies from event to event, and the overlapping events in polyphonic audio recordings are difficult to classify.

To use large-scale weakly labeled data for audio tagging, the convolutional recurrent neural network (CRNN) [1] has been used and has achieved state-of-the-art performance in audio tagging. To begin, the waveform of audio recording is transformed to time–frequency (T–F) representation such as the log-mel spectrogram. Then, convolutional layers are applied on the T–F representation to extract high-level acoustic features. This means the T–F representation of the audio recording is treated as an image, which is fed into the convolutional neural network (CNN). However, unlike image classification, where the objects are usually centered and occupy a dominant part of the image, sound events in real life may only occur a short time in an audio clip. To better recognize the sound events in audio clips, some attention mechanisms [6–8] for CNN are applied to attend to the sound events and ignore the background audio events. Attention is a learning technique inspired by human visual perception, which was first proposed in computer vision [8]. Attention could help the network automatically learn to describe the content of input features and steer the model to information important to the task to obtain better performance. Among these newly proposed attention mechanisms in CNN [6–8], the gated linear unit (GLU) [6] and the Squeeze-and-Excitation (SE) block [7] have been proposed with different concerns. The GLU introduces the attention mechanism to all the layers of CNN, which may learn to control the amount of information of each T–F unit flow to the next layer by the sigmoid function. The GLU focuses on the local information of the feature map in each channel; on the other hand, the SE block focuses on the channel-wise relationship of the feature maps. The feature maps are reweighted to generate the output of the SE block, which can then be fed directly into subsequent layers. To evaluate the performance of the two different attention mechanisms, we propose to use a CRNN with a SE block (SE-CRNN) and a CRNN with a GLU (GLU-CRNN) for weakly labeled audio tagging.

Two contributions are in this paper. First, in weakly labeled audio tagging, we apply the SE block in CRNN. Second, we compare the differences between the two attention mechanisms: the SE block and GLU. In addition, we analyze the reasons for the difference in performance of the SE-CRNN and GLU-CRNN in weakly labeled audio tagging. This paper is organized as follows: Sect. 2 introduces related works. Section 3 describes the model structure. Section 4 describes the dataset, experimental setup, evaluation metrics, and results. Section 5 gives conclusions.

2 Related Work

Most of the previous work in audio tagging use features such as mel-frequency cepstral coefficients (MFCC) and classifiers such as Hidden Markov Models (HMM) [9] or Gaussian Mixture Models (GMM) [2]. Recent audio tagging methods are widely based on deep neural networks (DNNs) [4] with inputs varying from short-time Fourier transform, log Mel energy, and MFCC to constant-Q transform (CQT) [10]. Inspired by image classification [11], CNN was applied on the spectrogram of audio clips [12]. In [12], the audio clip is treated as an image. Unlike the object in the image, in audio clips from real life, some events, such as "people speaking," may last a few minutes, while other events, such as "gunshot," may only last for hundreds of milliseconds. This characteristic of sound events increases the difficulty of audio tagging based on CNN.

To better use CNN to extract high-level representations of input features, attention [6, 7, 13] has been proposed in CNN. Attention can be viewed, broadly, as a tool to bias the allocation of available processing resources towards the most informative components of an input signal. The benefits of such a mechanism have been shown across a range of tasks, from understanding in images [13] and audio tagging [14] to language modeling [6]. Different from this paper, [14] explores the performance of attention scheme in recurrent neural network (RNN) layers for audio tagging, which shows that the attention method can alleviate the overfitting problem, especially when the input is chunk-level features.

Among the attention mechanisms, both the GLU and the SE block are based on the gates, which reduce the vanishing gradient problem by having linear units coupled to the gates (e.g., a sigmoid or softmax function). In this paper, we will explore the difference between the attention based on a GLU and the attention based on a SE block in weakly labeled audio tagging.

3 Model with Attention Mechanism

In this section, the CRNN baseline, CRNN with SE block (SE-CRNN) attention and CRNN with GLU (GLU-CRNN) attention will be introduced, respectively.

3.1 CRNN Baseline

CRNN has been successfully used in audio tagging tasks [1]. First, the waveforms of audio clips are transformed to T–F representations, such as log-mel spectrograms. Convolutional layers are applied on the T–F representations to extract high-level features. Then, bidirectional gated recurrent units (BGRUs) are adopted to capture the temporal context information. The BGRU layer is followed by a dense layer to predict the posteriors of each audio class at each frame. The activation function of the dense layer is sigmoid since audio tagging is a multiclass classification problem [1]. Finally, the prediction probability of each audio class label is obtained by averaging the posteriors of all the frames. Inspired by the good performance of CRNN in audio tagging, CRNN is used as the baseline system in this paper, as shown in Fig. 1.

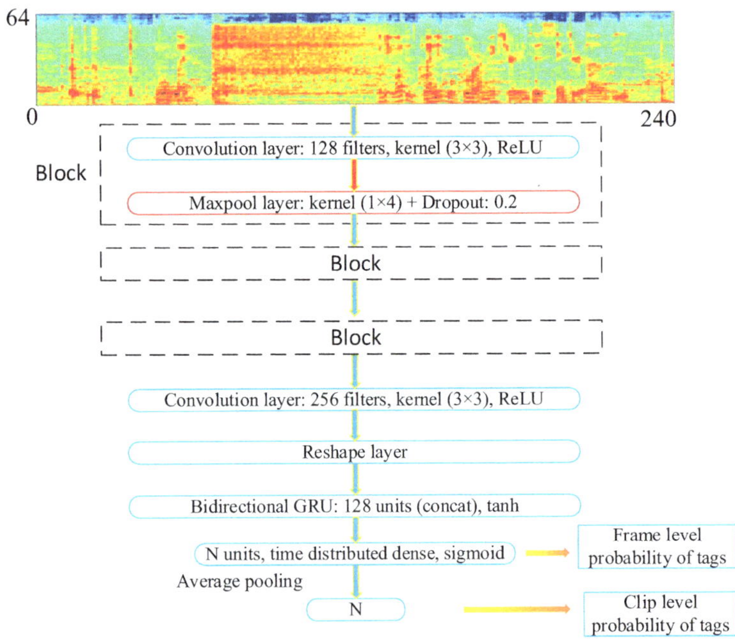

Fig. 1 CRNN model structure. *N* denotes the number of sound event classes. Other parameters will be described in later experiments setup

In Fig. 1, pooling is applied to the frequency axis only to preserve the time resolution of the input. The predicted clip level probability of tags can be obtained from the last layer. Binary cross-entropy is used as loss between the predicted probability and the ground truth of an audio recording, since it was demonstrated to be better than the mean squared error in [15] for labels with zeros or one value. The loss can be defined as

$$E = -\sum_{m=1}^{M}(\boldsymbol{R}_m \log \boldsymbol{P}_m + (1 - \boldsymbol{R}_m)\log(1 - \boldsymbol{P}_m)) \tag{1}$$

where E is the binary cross-entropy, and $\boldsymbol{P}_m$ and $\boldsymbol{R}_m$ denote the estimated and reference ground truth tag vector at sample index m, respectively. The batch size is represented by M.

3.2 *Model with SE Attention: SE-CRNN*

To find which channel of the feature maps is more important, we propose to use the attention of SE blocks to all the convolution layers in the CRNN for audio tagging. The SE block [7] is a new attention mechanism in CNN, which focuses on the channel relationship in convolution operation in CNN and adaptively recalibrates channel-wise feature responses by explicit interdependencies between channels. The basic structure of a SE block is illustrated in Fig. 2.

Given X is input data and U is features extracted from X, then transformation $\mathrm{F_{tr}}$: $\mathrm{X} \rightarrow \mathrm{U}$, $\mathrm{X} \in \mathcal{R}^{H' \times W' \times C'}$, $\mathrm{U} \in \mathcal{R}^{H \times W \times C}$, where the H and W are the height and width of feature maps, and C denotes the channel number. In convolution layers, the SE block can be constructed to perform the feature recalibration. The features U are passed through a *squeeze* operation [7], which aggregates the feature maps across dimensions $H \times W$ to produce a channel descriptor. The descriptor embeds the global distribution of channel-wise feature responses so that the information from the global receptive field of the network will be leveraged by the lower layers. Next is the *excitation* operation [7], in which sample-specific activations are learned for each channel by a self-gating mechanism based on channel dependence. The feature

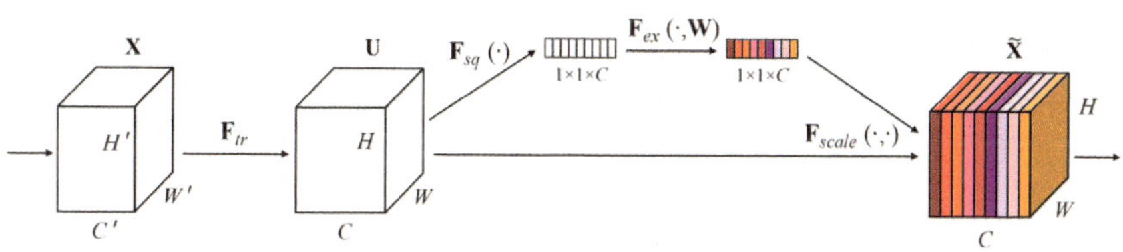

Fig. 2 A Squeeze-and-Excitation (SE) block [7]

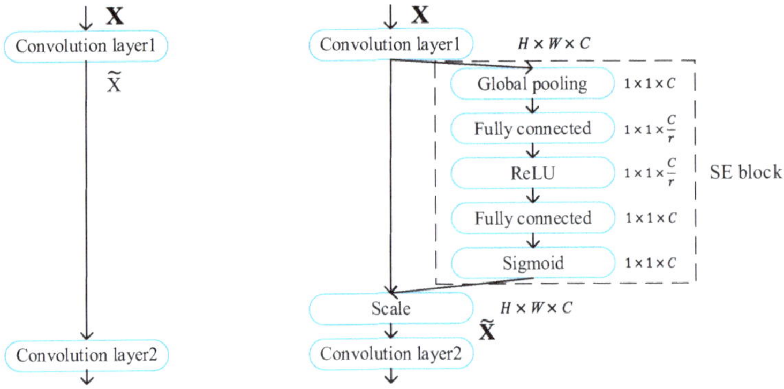

Fig. 3 The original convolution layer (left) and the SE convolution layer (right)

maps U are then reweighted to generate the output of the SE block. More details about the SE block can be seen in [7].

The flexibility of the SE block means that it can be directly applied to the convolution layers, as shown in Fig. 3. The reduction ratio r is an important hyperparameter which allows us to vary the capacity and computational cost of the SE block. We will investigate r in Sect. 4.2.

3.3 Model with GLU Attention: GLU-CRNN

Unlike the SE block, which focuses on the channel-wise relationship of the feature maps, the GLU attention focuses on the local information of the feature map in each channel. Compared with SE-CRNN, GLU-CRNN applies the attention mechanism on the T–F representations of each feature map. That is, each T–F unit of a feature map is attended differently. The GLU is proposed in [6] in language modeling, and the structure of the GLU is shown in Fig. 4. By providing a linear path for the propagation of the gradients while keeping nonlinear capabilities through the sigmoid operation, the GLU can reduce the gradient vanishing problem in deep networks [6], and the GLU can control the amount of information of a T–F unit flow to the next layer by the sigmoid function. Given W and V are convolutional filters, b and c are biases, X denotes the input T–F representations in the first layer or the feature maps of the interval layers, and σ is sigmoid function, then GLU could be defined as

$$Y = (W * X + b) \odot \sigma(V * X + c) \tag{2}$$

where the symbol $\odot$ is the element-wise product and $*$ is the convolution operator. Another benefit of using a GLU is that the network can learn to attend to sound events and ignore the unrelated sounds. That is, if the value of the sigmoid function

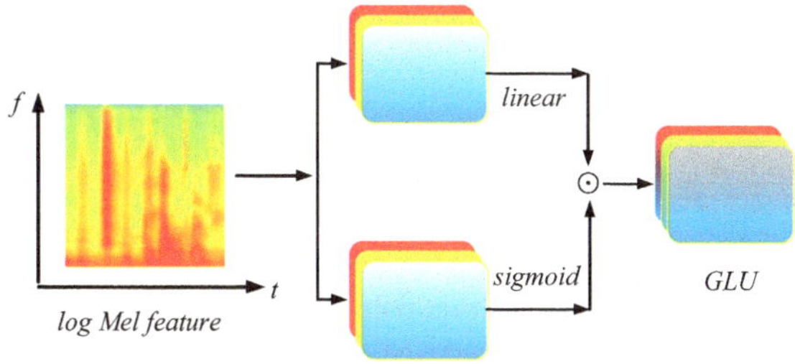

Fig. 4 The Structure of GLU

is close to 1, then the corresponding T–F unit is attended. If the value of the sigmoid function is near to 0, then the corresponding T–F unit is ignored.

4 Experiments and Results

4.1 Dataset, Experiment Setup and Evaluation Metrics

The audio tagging dataset used in this paper is polyphonic audio in real life from the DCASE 2018 challenge Task 4 labeled training set [16], which is a subset of the Google Audio Set. Audio Set consists of an ontology of 632 sound event classes and a collection of 2 million human-labeled 10-second audio clips drawn from YouTube. The dataset in the DCASE 2018 challenge Task 4 consists of 10 classes of sound events, including "Speech," "Dog," "Cat," "Alarm bell ringing," "Dishes," "Frying," "Blender," "Running water," "Vacuum cleaner," and "Electric shaver."

In the training phase, log mel-band energy is extracted using Short-time Fourier Transform (STFT) with a Hamming window with 64 mel-frequency bins [17]. For a given audio clip of 10 s, this feature extraction block results in a (240, 64) output with 240 frames and 64 mel-frequency bins (Fig. 1). Dropout and early stopping criteria are used to prevent overfitting. The model is trained for maximum 200 epochs with an Adam optimizer with a learning rate of 0.001.

To evaluate the results of audio tagging, the metrics of precision (P), recall (R), F-score [17], and area under curve (AUC) [18] need to be calculated. To calculate these metrics, we need to count the numbers of true positive (TP), false negative (FN), and false positive (FP). The P, R, and F-scores are defined as

$$P = TP/(TP + FP) \tag{3}$$

$$R = TP/(TP + FN) \tag{4}$$

$$F = (2P \cdot R)/(P + R) \tag{5}$$

To compare the true positive rate (TPR) versus the false positive rate (FPR), the receiver operating characteristic (ROC) curve was used, and the AUC score is the area under the ROC curve [18]. Larger P, R, and F-scores and AUC indicate better performance.

4.2 Result and Analysis

The main purpose of this paper is to compare the performance of the two different attention mechanisms, the SE block and GLU, on CNN learning power of high-level representations in audio tagging. Thus, we propose to use SE-CRNN and GLU-CRNN in weakly labeled audio tagging. To obtain better channel-wise attention information in SE-CRNN, the reduction ratio r values [7] were investigated in Table 1; various values of r are on the x axis, and the metrics score is on the y axis. The comparison in Table 1 reveals that performance does not improve monotonically with increased capacity. The reason may be that the SE block is overfitting the channel interdependencies of the training set. In particular, we found setting $r = 16$ achieved a good trade-off between accuracy and complexity. Consequently, we used this value for all experiments in the SE-CRNN model.

To comprehensively compare the performance of the two different attention mechanisms, we calculated the P, R, and F-scores, and AUC of different models. The averaged statistic in Table 2 shows the GLU-CRNN performs better than the SE-CRNN and the baseline CRNN model overall.

The overall performance of each model is shown in Table 2. In order to deeper investigate the difference of attention mechanisms on each class of sound events, we calculated the AUC over 10 kinds of sound events in this task. As shown in Table 3,

Table 1 Different reduction ratios r in SE-CRNN

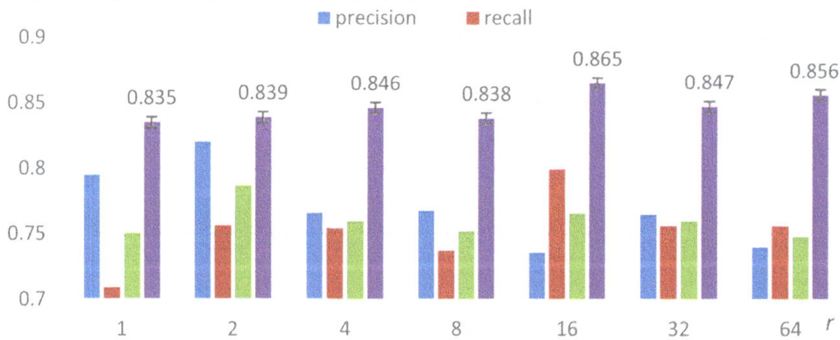

Table 2 Averaged stats of audio tagging

	Precision	Recall	F-score	AUC
Baseline	0.759	0.752	0.755	0.838
SE-CRNN	0.736	**0.799**	0.766	0.865
GLU-CRNN	**0.855**	0.797	**0.825**	**0.877**

Table 3 AUC of audio tagging over 10 kinds sound events

	Speech	Dog	Cat	Bell	Dishes	Frying	Blender	Water	Cleaner	Shaver	Avg.
Baseline	0.906	0.963	0.960	0.931	0.723	0.827	0.583	0.840	0.875	0.774	0.838
SE-CRNN	0.912	0.969	0.949	0.929	**0.772**	0.818	0.767	0.804	0.883	**0.848**	0.865
GLU-CRNN	0.906	0.965	0.951	**0.963**	0.732	0.799	**0.824**	**0.890**	**0.940**	0.796	**0.877**

for the audio classes including "Speech," "Dog," and "Cat", the performances of SE-CRNN and GLU-CRNN are similar. However, for the audio classes "Dishes" and "Shaver," the AUC of the SE-CRNN is higher. For the audio classes such as "Bell," "Blender," "Water," and "Cleaner," the AUC of the GLU-CRNN is higher, and its overall performance is better than the SE-CRNN and the baseline CRNN model. The reason may be that the attention based on the GLU is better at capturing the sound events with a more pronounced rhythm like "Bell," "Blender," and "Cleaner." To further analyze the difference between the two attention mechanisms, Fig. 5 shows the internal high-level representations of one feature map, which was randomly selected from the feature maps learned by the convolution layer of each model.

For subfigures (b), (c), and (d) in Fig. 5, the frame is on the x axis, and the channel of convolution layer is on the y axis. Compared with the internal output of the baseline CRNN, the internal output of the SE-CRNN retains the information of the former channels and deletes the information of the latter channels. The reason may be after adding the channel-wise attention mechanism of the SE block, the SE-CRNN model considers the information of some channels to be more useful, thus retaining the information of the useful channels and deleting the information of the useless channels. For the internal output of GLU-CRNN, the attention based on GLU focuses on the local information within each feature map, not the relationship between the channels. The GLU model can weight T–F units separately according to their unique time positions, at which sound events may or may not occur. In contrast to the GLU, the SE model weights the T–F units in a channel as a whole, ignoring the different time positions of the units in the channel, which is obviously contrary

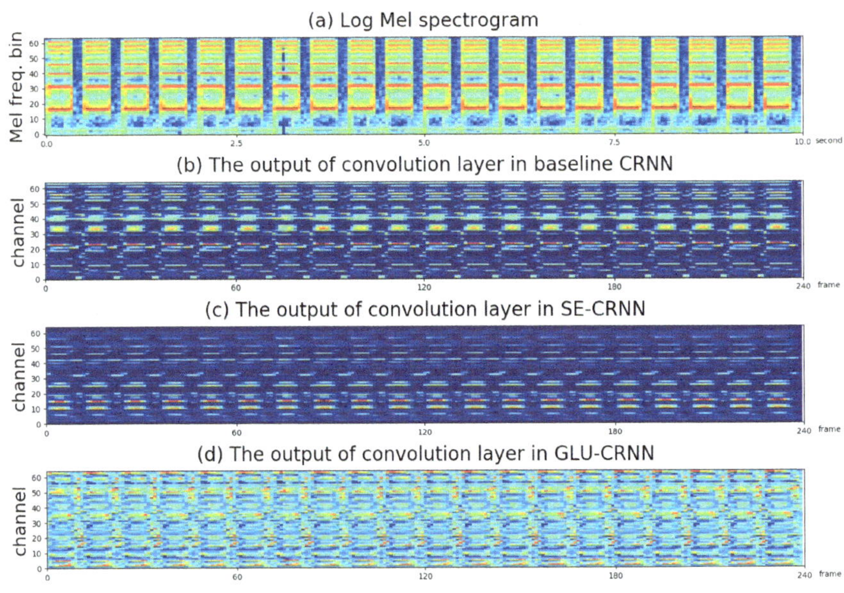

Fig. 5 The internal high-level representations of one feature map

to the idea of discriminating T–F units by the presence and absence of events at their time positions. This may be the reason why the GLU performs better than SE. As a result, the GLU-CRNN learns the high-level representations of local information within each feature map better, and the overall performance of the GLU-CRNN in the experiments is the best.

5 Conclusion

In this paper, we compare the performance of the two different attention mechanisms, SE block and GLU, based on SE-CRNN and GLU-CRNN in clip-level audio tagging, respectively. The experiments show the attention of GLU is better than the attention of SE in weakly labeled audio tagging. The reason may be the GLU can weight T–F units separately according to their unique time positions, at which sound events may or may not occur. However, SE weights the T–F units in a channel as a whole, ignoring the different time positions of the units in the channel. Future work will focus on comparing the difference between the GLU and SE in sound event detection.

Acknowledgement Qiuqiang Kong was supported by the China Scholarship Council (CSC) No. 201406150082.

References

1. Xu Y, Kong Q, Wang W, Plumbley MD (2018) Large-scale weakly supervised audio classification using gatedconvolutional neural network. In: IEEE International Conference on Acoustics, Speech and Signal Processing (ICASSP) 2018, Calgary Canada, 2018, (pp 121–125)
2. Stowell D, Giannoulis D, Benetos E, Lagrange M, Plumbley MD (2015) Detection and classification of acoustic scenes and events. IEEE Trans Multimed 17(10):1733–1746
3. Dimitrov S, Britz J, Brandherm B, Frey J (2014) Analyzing sounds of home environment for device recognition. In: European Conference on Ambient Intelligence, pp 1–16
4. Kumar A, Raj B (2016) Audio event detection using weakly labeled data. In: ACM on Multimedia Conference, pp 1038–1047
5. Mesaros A, Heittola T, Diment A, Elizalde B, Shah A, Vincent E, Raj B, Virtanen T (2017) DCASE 2017 challenge setup: Tasks, datasets and baseline system. In: Workshop on Detection and Classification of Acoustic Scenes and Events (DCASE) 2017, Munich, Germany, 2017
6. Dauphin YN, Fan A, Auli M, Grangier D (2017) Language modeling with gated convolutional networks. In: Proceedings of International Conference on Machine Learning (ICML), 2017, pp 933–941
7. Hu J, Shen L, Sun G (2018) Squeeze-and-excitation networks. In: Proceedings of the IEEE conference on computer vision and pattern recognition, pp 7132–7141
8. Mnih V, Heess N, Graves A (2014) Recurrent models of visual attention. In: Advances in neural information processing systems, pp 2204–2212
9. Mesaros A, Heittola T, Eronen A, Virtanen T (2010) Acoustic event detection in real life recordings. In: European Signal Processing Conference. IEEE, pp 1267–1271
10. Lidy T, Schindler A (2016) CQT-based convolutional neural networks for audio scene classification: In: Proceedings of the detection and classification of acoustic scenes and events 2016 workshop 90:1032–1048

11. Krizhevsky A, Sutskever I, Hinton GE (2010) Imagenet classification with deep convolutional neural networks. In: Advances in neural information processing systems, pp 1097–1105
12. Choi K, Fazekas G, Sandler M (2016) Automatic tagging using deep convolutional neural networks. In: arXiv preprint, arXiv:1606.00298
13. Itti L, Koch C, Niebur E (1998) A model of saliency-based visual attention for rapid scene analysis. In: IEEE Transactions on Pattern Analysis and Machine Intelligence, pp 1254–1259
14. Xu Y, Kong Q, Huang Q, Wang W, Plumbley MD (2017) Attention and localization based on a deep convolutional recurrent model for weakly supervised audio tagging. In: arXiv preprint, arXiv:1703.06052
15. Xu Y, Huang Q, Wang W, Foster P, Sigtia S, Jackson PJ, Plumbley MD (2017) Unsupervised feature learning based on deep models for environmental audio tagging. In: IEEE/ACM Transactions Audio, Speech, Language Process 25(6):1230–1241
16. Serizel R, Turpault N, Eghbal-Zadeh H (2018) Large-scale weakly labeled semi-supervised sound event detection in domestic environments. In Workshop on Detection and Classification of Acoustic Scenes and Events (DCASE) 2018, November 2018, Surrey, UK
17. Mesaros A, Heittola T, Virtanen T (2016) Metrics for polyphonic sound event detection. Appl Sci 6(6):162
18. Hanley JA, McNeil BJ (1982) The meaning and use of the area under a receiver operating characteristic (ROC) curve. Radiology 143(1):29–36

Music Steganography

A Standard MIDI File Steganography Based on Music Perception in Note Duration

Lei Guan, Yinji Jing, Shengchen Li and Ru Zhang

Abstract Steganography aims to deliver messages in the recording space of a cover media without being noticed. Supported by the theory of music perception, note duration can be used as the recording space for a steganography method. This paper proposes a steganography method that embeds information in the duration of notes in a MIDI file and generates a new MIDI file in which the only difference lies in the note duration. Under several restrictions, the available varying range of each note is calculated, and then secret information is mapped on the change in note duration within the available range. A listening test and an evaluation on capacity are conducted to measure the transparency and capacity of the proposed method. The results reveal that the proposed method has perfect transparency with no perceptible difference in hearing, zero expansion in file size, and an effective capacity of 3.5%.

Keywords Steganography · MIDI · Note duration

1 Introduction

Steganography is the science of communicating secret information in a hidden manner [1]. The general principle for steganography is to embed information in a digital cover media that results in a Stego-media. The place where information hides in the cover media is called the recording space, which is crucial to the steganography

L. Guan · Y. Jing · S. Li (✉) · R. Zhang
Beijing University of Posts and Telecommunications, No. 10 Xitucheng Road, Haidian District, 100876 Beijing, China
e-mail: shengchen.li@bupt.edu.cn

L. Guan
e-mail: guanlei@bupt.edu.cn

Y. Jing
e-mail: jyj@bupt.edu.cn

R. Zhang
e-mail: zhangru@bupt.edu.cn

W. Li et al. (eds.), *Proceedings of the 6th Conference on Sound and Music Technology (CSMT)*, Lecture Notes in Electrical Engineering 568,
https://doi.org/10.1007/978-981-13-8707-4_9

method. There are many kinds of conversional recording space [2], such as open space in text media, which has high capacity but poor transparency, least-significant bits in graphics, which are extremely vulnerable, and so on. The proposed method takes advantage of note duration in music as the recording space, which does not have the shortcomings of the methods mentioned above.

The JND (Just Noticeable Difference) of music perception of note duration is the main supportive theory for the proposed method. As the ability of the human auditory system is limited, differences in note duration as well as other aspects are not perceptible on a certain scale. The JND describes the largest difference under the threshold of human perception. According to [3], the JND of note duration is about 10% of the duration. Dahl and Granqvist [4] reported that the JND of tempo changes is about 2% of the tempo. A later report [5] increases the JND of tempo changes to 8%, but the specific JNDs in different circumstances are still unclear. The proposed method changes the duration of notes within a JND threshold value of 8% to embed data. We chose 8% to ensure transparency. However, larger values can be used to further increase the capacity of the proposed system.

Coding information in note duration requires a precise representation of note duration. Famous for accurate timing representation with compact file size, MIDI is one of the commonly used cover media in steganography. MIDI is a standard protocol for exchanging music information between electronic musical instruments and computers that dates back to the early 1980s [6]. A standard MIDI file is formed by chunks containing a chunk tag, the chunk length, and the chunk data. There are two types of chunks in MIDI: Header Chunks and Tracker Chunks. The Header Chunk contains the MIDI file's general information, such as the total bytes in the file, the file format, and the number of tracks. Tracker Chunks consist of events and delta-time, where "events" refers to operation instructions for music players, including MIDI control events, system exclusive events, and meta events, and "delta-time" refers to the time interval between the current event and the previous event and is used to precisely represent the time of each event. Specifically speaking, the happening time of each event is the sum of its delta-time and that of all the previous events.

The meta data of meta events such as author and album information provide a recording space to store secret information. Malcolm [7], Inoue and Matsumoto [8], and Adli and Nakao [9] used this method to embed information in MIDI files. Such methods have a high capacity for information but their transparency is poor, i.e., the file size of the Stego-media is increased. Besides meta data, velocity and note duration are also used to hide information. Adli and Nakao [9], Adli [10], and Wu and Chen [11] proposed steganography algorithms based on velocity values, which have good transparency but limited capacity. Adli [10], Yamamoto and Iwakiri [12], and Xu et al. [13] tried to change the duration of notes to embed information in MIDI files. They achieved good capacity but had some expansion of file size.

The proposed method embeds information in note duration as well. It has an equivalent capacity but better transparency with no perceptible difference in hearing and zero expansion of file size compared to other steganography methods using note duration. We conduct a listening test to verify whether the duration changes caused by the proposed method are perceptible. We conduct an evaluation of capacity as well.

The paper is organized as follows. First, we introduce the fundamental theory of music perception of note duration and other related background knowledge. Next, we describe the implementation of the proposed method, followed by the performance measurements of the proposed method compared to related work. The paper ends with a summary of our findings.

2 System Implementation

The proposed method embeds information in note duration, that is, it codes information to changes in note duration. As MIDI is the cover media, we introduce our method's necessary operation on MIDI in detail. To embed information, we first evaluate the capacity of each note according to the information of the whole MIDI file. Next, we map the secret information to changes in note duration. Finally, we generate a new MIDI file containing the changed note durations. The process of information extracting reverses the process of information embedding.

2.1 Operation on MIDI

The only modification needed is to the note duration, which is represented by the time difference between the note-on event and note-off event of the same note. To keep the melody of the music constant, the start time of each note cannot be modified, that is, the time of the note-on event. Therefore, the time of the note-off event is the only space left to change the note duration. As introduced above, the time of each event, including the note-off event, is represented by the sum of its delta-time and the delta-time of all previous events. Changing the delta-time of the note-off event will change its time and the duration of the note.

Figure 1 presents a simplified example of a note in MIDI. Each event is recorded with four bytes. The first of these, which is the event's delta-time, is what we focus on. There are no other events between the note-on event and the note-off event, so the note duration equals the delta-time of the note-off event, which is to be modified.

To keep the size of the MIDI file constant, any modification cannot result in a greater or lesser number of bytes for storage. The only modification is to the delta-time of the note-off event. Delta-time is recorded with a method called dynamic byte, which uses different numbers of bytes to record different values of delta-time. From the first byte on, the highest bit "1" of the byte means there is at least one more byte following as the next part of the delta-time, and the highest bit "0" means this byte is the last byte of the delta-time. For each byte, seven lower bits are valid for recording values. If there is more than one byte, the seven lower bits of the first byte record within a range of 1–127 (zero means the byte is redundant), while the seven lower

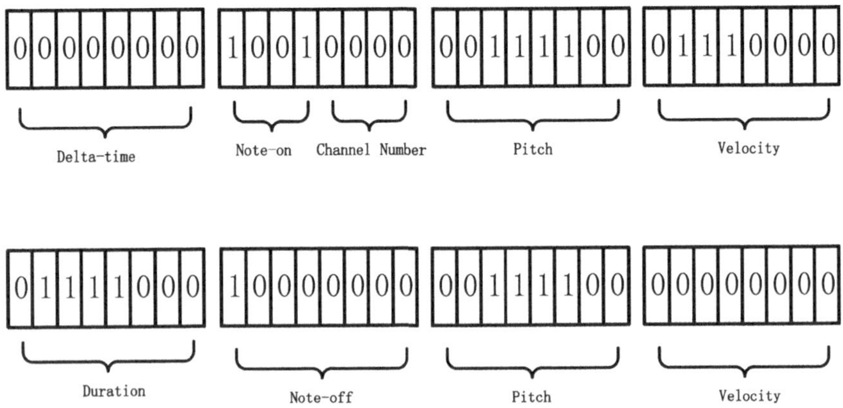

Fig. 1 The representation of a note

bits of other bytes record within a range of 0–127. And if there is only one byte, like the delta-time of the note-off event in Fig. 1, the available range is 0–127.

For the delta-time of k bytes, the minimum value t_{min} is

$$t_{min} = \begin{cases} 128^{k-1}, & k > 1 \\ 0, & k = 1 \end{cases}, \tag{1}$$

and the maximum value t_{max} is

$$t_{max} = 128^{k} - 1. \tag{2}$$

Modification of delta-time is allowed only within the range $[t_{min}, t_{max}]$. If the corresponding range of the change in note duration is different from the range of the JND threshold or other ranges, the smallest is chosen, which we introduce as follows.

2.2 Procedure of Information Embedding

With given hidden binary information and a given MIDI file as the cover media, a new MIDI file, the Stego-media, is generated. The two MIDI files should have the same file size and only an imperceptible difference in note duration. For example, for a certain note, the original duration is represented by L, the new duration by L', and the change of duration by $\triangle L$. The capacity C of the note denotes the number of bits that can be embedded. The C bits information is read and mapped to the duration change $\triangle L$. After the new duration of each note is calculated, the new MIDI file is generated. The detailed three steps are as follows.

Step 1. Evaluating the capacity of each note. For the available range $[-R, R]$ of the duration change ΔL, there are two constraints: R_1 on the auditory perception threshold and R_2 on the MIDI format.

Using the JND theory [8], we chose 8% of the note as the threshold, a choice that has been supported by the results of later experiments. With the note duration L, the R_1 is

$$R_1 = L * 8\%. \tag{3}$$

The constraint R_2 includes keeping events in the original order and keeping the total time of the music and the MIDI file size constant. To keep events in the original order, for any event the changed time cannot occur before the last event or after the next event. To keep the total time of the music constant, the time of each event cannot occur after the original end time of the music. To keep the MIDI file size constant, the proposed method changes the time in such a range that the ticks in the Stego-MIDI do not need a greater or lesser number of bytes for storage. Therefore, the expansion of the file is zero.

With R_2 obtained, the overall constraint R is the minimum of R_1 and R_2

$$R = min\{R_1, R_2\}. \tag{4}$$

The capacity C is then calculated by

$$C = \lfloor log_2(R + 1) \rfloor. \tag{5}$$

Step 2. Mapping binary information to the change in note duration. For each note with capacity C, the C-bit information is calculated to a decimal value a. Both C and a need to be coded because the C-bit information can start with several zeros that cannot be extracted using only a. The duration change ΔL is then designed as

$$\Delta L = \begin{cases} a, & a \geq 2^{C-1} \\ (-1) * \left(a + 2^{C-1}\right), & a < 2^{C-1} \end{cases}. \tag{6}$$

With such a mapping method, both C and a can be extracted from ΔL, as we demonstrate later.

Step 3. Embedding the information into MIDI files. For each note, the new duration L' is the sum of the original duration L and the duration change ΔL. To generate the new MIDI file, the delta-time of each event is recalculated using the time difference between the event and the previous event to obtain the Stego-MIDI.

2.3 Procedure of Information Extraction

The procedure of information extraction reverses the procedure of information embedding. For each note, ΔL is obtained by calculating the difference between the note duration in the new MIDI and in the original MIDI. $|\Delta\ L|$ is defined as the absolute value of ΔL in the binary form. The capacity C equals the number of binary bits of $|\Delta\ L|$. a is calculated by

$$a = \begin{cases} \Delta\ L, & \Delta\ L > 0 \\ (-1) * \left(\Delta\ L - 2^{C-1}\right), & \Delta\ L < 0 \end{cases}. \tag{7}$$

The original C-bit information is a in the binary form with the rest bits in the front filled with zeros.

3 Experiments and Evaluations

Steganography methods are commonly measured by three aspects: transparency, capacity, and robustness. We conduct experiments and evaluations to measure transparency and capacity. Our findings demonstrate that the proposed method has better transparency than other methods and good capacity.

However, all steganography methods using MIDI have poor robustness because several bits of reverse can break the MIDI format. In other words, MIDI itself has a poor robustness for which steganography methods are not to blame.

3.1 Transparency

Transparency describes whether the proposed steganography method leaves any noticeable feature that is likely to expose the secret communication. The proposed method only modifies one parameter, the delta-time, of some events in the MIDI file. It does not damage of MIDI format nor expand the file size. Other steganography methods using note duration have some degree of file size expansion, thus the proposed method has better transparency at the file storage level.

We then conduct a listening test to verify transparency at the music perception level. Specifically, we design the listening test to verify that the duration changes made by the proposed method are not perceptible. Ten listeners received a package online that contained 10 pairs of MIDI files, half of which include a cover MIDI and a Stego-MIDI and half two same copies of a cover MIDI file. Without knowing if there is any difference between each pair of MIDI files, the listeners were asked to listen carefully to the 10 pairs using their own audio playback devices and pick out the pairs that had differences they could perceive.

Table 1 Results of listening test

Listeners	The same	Different	Correct	Correct rate (%)
Listener 1	8	2	5	50
Listener 2	10	0	5	50
Listener 3	10	0	5	50
Listener 4	6	4	5	50
Listener 5	10	0	5	50
Listener 6	10	0	5	50
Listener 7	9	1	4	40
Listener 8	10	0	5	50
Listener 9	3	7	4	40
Listener 10	10	0	5	50

The result of the listening test is shown in Table 1. "The Same" is the number of pairs of MIDI files that sound the same to the listener. "Different" is the number of pairs that sound different and were picked out by the listeners. "Correct" is the number of times the 10 judgments were correct. "Correct Rate" is the ratio of "Correct" and the total number of judgments.

In all ten pairs of MIDI files, half of them were the same and half different. If a listener randomly picks out the differing pairs, the expectation for the Correct Rate is 50%. An actual Correct Rate that is equal to or lower than 50% indicates that the listener cannot perceive the difference. As the table shows, none of the listeners had a Correct Rate higher than 50%; therefore none could perceive the differences. These results show that the proposed method has good transparency at the music perception level. In conclusion, the proposed method has better transparency than other steganography methods using note duration.

3.2 Capacity

Capacity, which represents how much information can be embedded in the cover media, is measured by the ratio of embedded bits and the size of the bits of the MIDI file, shown in Table 2. We embedded information in five MIDI files randomly collected from the internet. For these files, "File Size" is the size in bytes of the MIDI file. "Embedded Bits" is the number of bits of secret information embedded in the MIDI file. "Capacity" is defined above. The maximum capacity of the proposed method is 3.50%.

Table 3 shows the evaluation results of capacity for our proposed method compared to other methods. "Event" shows the result with Stegokey for eight chords of the Inoue et al. [14] method, "Velocity" shows the evaluation result of the Yamamoto et al. [15] method, and "Fluctuation" shows the evaluation result of the Yamamoto et al. [12]

Table 2 Capacity of MIDI files

	File size [byte]	Embedded bits [bit]	Capacity (%)
MIDI 1	774	212	3.42
MIDI 2	5126	1217	2.97
MIDI 3	4068	827	2.54
MIDI 4	1328	372	3.50
MIDI 5	2328	427	2.29

Table 3 Capacity of different steganography methods

	Event	Velocity	Fluctuation	Proposed
Maximum (%)	4.01	3.58	5.55	3.50

method. The capacity of the proposed method is slightly lower than that of the other methods but basically at the same level. What is more, since the proposed method is compatible with the other methods, a larger capacity can be achieved by combining the proposed method with the others.

4 Conclusion

In this paper, we proposed a standard MIDI file steganography based on music perception of note duration. Supported by the theory of music perception, we mapped secret information to the changes in note duration and then embedded it in a MIDI file. Our results show that the proposed method has better transparency than other steganography methods using note duration and a good capacity of 3.5%. With its good transparency, the proposed method is suitable for confidential correspondence. For instance, it is possible to embed the proposed method in electronic musical instruments to automatically generate information-embedded music.

In addition, if needed, the proposed method can be combined with other MIDI steganography methods and use a higher JND threshold to increase embedding capacity.

References

1. Cheddad A., Condell J, Curran K, Kevitt PM (2010) Digital image steganography: Survey and analysis of current methods. Signal Process 90(3):727–752
2. Thampi SM (2008) Information hiding techniques: a tutorial review. arXiv preprint arXiv: 0802.3746
3. Stevens SS (1951) Handbook of experimental psychology. J Am Med Assoc 148(1):117

4. Dahl S, Granqvist S (2003) Looking at perception of continuous tempo drift: a new method for estimating internal drift and just noticeable difference. In: Proceedings of the Stockholm Music Acoustics Conference (SMAC 03), vol 2, pp 595–598
5. Thomas K (2007) Just noticeable difference and tempo change. J Sci Psychol 2:14–20
6. Midi manufacturer association. https://www.midi.org/, last assessed 15 Sept 2018
7. Malcolm JW (2004) Method and apparatus for encoding security information in a midi datastream (Sep 28 2004), US Patent 6,798,885
8. Inoue D, Matsumoto, T (2002) Scheme of standard midi files steganography and its evaluation. In: Security and Watermarking of Multimedia Contents IV. vol 4675, International Society for Optics and Photonics, pp 194–206
9. Adli N (2005) Three steganography algorithms for midi files. In: International conference on machine learning and cybernetics, vol 4, pp 2401–2404
10. Adli A, Mirza H, Nakao Z (2008) A watermarking approach for midi file based on velocity and duration modulation. In: International conference on knowledge-based and intelligent information and engineering systems, pp 133–140
11. Wu DC, Chen MY (2016) Information hiding in standard midi files based on velocity reference values. IJ Network Sec 18(2):274–282
12. Kotaro Y, Munetoshi I (2009) A standard midi file steganography based on fluctuation of duration. In: | 2009 International conference on availability, reliability and security, pp 774–779. IEEE
13. Xu C, Zhu Y, Feng DD (2001) Content protection and usage control for digital music. In: Proceeding of international conference on web delivering of music, pp 43–50
14. Inoue D (2002) Standard midi files steganography and its power. IPSJ Trans 43(8):2489–2501. https://ci.nii.ac.jp/naid/10011451492/en/
15. Yamamoto K, Iwakiri M (2006) A steganography to music code with adaptation in musical expression. Trans Inform Process Soc Jpn 47(8):2